Colour Change in Paintings

Colour Change in Paintings

Edited by
Rhiannon Clarricoates, Helen Dowding
and Alexandra Gent

Archetype
Publications

in association with
the Icon Paintings Group

THE INSTITUTE OF CONSERVATION

First published 2016 by Archetype Publications Ltd in association with the Icon Paintings Group

Archetype Publications Ltd
c/o International Academic Projects
1 Birdcage Walk
London SW1H 9JJ
www.archetype.co.uk

ISBN: 978-1-909492-43-1

British Library Cataloguing in Publication Data
A catalogue record for this book is available from the British Library.

Front cover illustration: Detail from Cy Twombly's painting *Quattro Stagioni. Part I: Primavera*, 1993–94 (© Cy Twombly Foundation).

Printed on acid-free paper

Designed by Marcus Nichols at PDQ Digital Media Solutions Ltd.
Typeset by PDQ Digital Media Solutions Ltd, Bungay
Printed and bound in Great Britain by Latimer Trend & Co Ltd, Plymouth

Contents

Foreword

As painting conservators, we accept the changes in colour that have occurred due to the chemical alteration of pigments and try to reduce the occurrence and extent of these changes through the control of light levels and environmental parameters. The ethical concerns of all those responsible for cultural heritage must include awareness of the effects of light and ageing on colours.

Artists typically choose their colours with care and consideration for their specific visual impact on the viewer. Painting conservators remove years of dirt and discoloured varnishes to ensure that those carefully chosen tones can be seen as close to the artist's intention as possible. It therefore makes sense that conservators should be concerned with understanding the shifts in colour and tone that can occur in paintings over time.

The papers in this volume were presented at the Icon Paintings Group conference *Appearance and Reality: Examining Colour Change in Paintings* held at Tate Britain on 9 October 2015. The conference offered an opportunity to hear from international speakers who have spent time researching and measuring colour change and its effects, and attempted to present an overview of the current state of research related to colour change in paintings, ranging from Old Masters to modern art. The contributors addressed not only analytical investigation into changes in painting materials, but also methods of assessing colour change, lighting paintings, alternative methods of display, and different approaches to the restoration of paintings whose colour and tonality have shifted.

We hope that this publication will be of interest to all those involved in the care and restoration of paintings.

Francis Downing ACR
Chair of the Icon Paintings Group Committee

Acknowledgements

We are extremely grateful to the contributing authors for their presentations and posters at the conference, as well as agreeing to publish their research, without which this book would not be possible.

The editorial sub-committee would like to emphasise the work of the whole of the Icon Paintings Group Committee in organising and delivering the conference.

Finally, the committee would like to acknowledge Tate for its generous support and for providing an excellent venue for the conference and evening reception. We would like to thank all Tate's staff who made this possible and provided assistance on the day.

Fading, darkening, browning, blanching: a review of our current understanding of colour change and its consequences in Old Master paintings

Marika Spring

Introduction

In many Old Master paintings most of the colours are remarkably well preserved, even after hundreds of years, but quite often some degradation of less stable pigments has occurred, altering the appearance and perception of a work. Red and yellow lake pigments based on organic dyestuffs that are sensitive to light can fade, as can indigo. The blue cobalt-containing glass pigment smalt can lose its colour, turning the paint brown or grey-brown, vermilion can blacken and paints based on verdigris that were originally green can become brown. Less obviously, the arsenic sulphide pigments, orpiment and realgar, can become lighter, as can red lead, and rather more subtle changes can result from the blanching of paint containing ultramarine or green earth. An awareness and knowledge of these degradation phenomena is important when cleaning and restoring paintings, and informs other aspects of conservation such as preventive measures through environmental control.

Artists have always understood that certain pigments can change in colour or deteriorate in other ways, as is evident from historic documentary sources, where measures intended to ameliorate and limit the effects of this lack of permanence can be found. Fugitive red lakes such as brazilwood are stated to be prohibited for use in high quality works of art (Kirby *et al.* 2006), we are told that orpiment should not be mixed with other pigments because it would cause them to darken (van Eikema-Hommes 2004: 36), and we are instructed to use indigo only in underpaint because of its tendency to fade (van Eikema-Hommes 2004: 133–6). This advice was based on the empirical observations and experience of artists over the ages, but in more recent times we have developed the scientific means to investigate in great depth the chemical mechanisms involved in pigment deteriorations. We can gain far more understanding of the effect of environmental factors such as light, humidity and pollution, and can investigate more effectively how the way in which a pigment has been used or prepared has influenced its stability. Much scientific research has been carried out on the causes, mechanisms and consequences of pigment alterations, especially in recent years. This paper reviews our current understanding of the changes most commonly encountered in Old Master paintings and considers how this knowledge can help in understanding their effect on what is seen on paintings. The research in this field is extensive and cannot be comprehensively covered here, so this review will concentrate mainly on work on

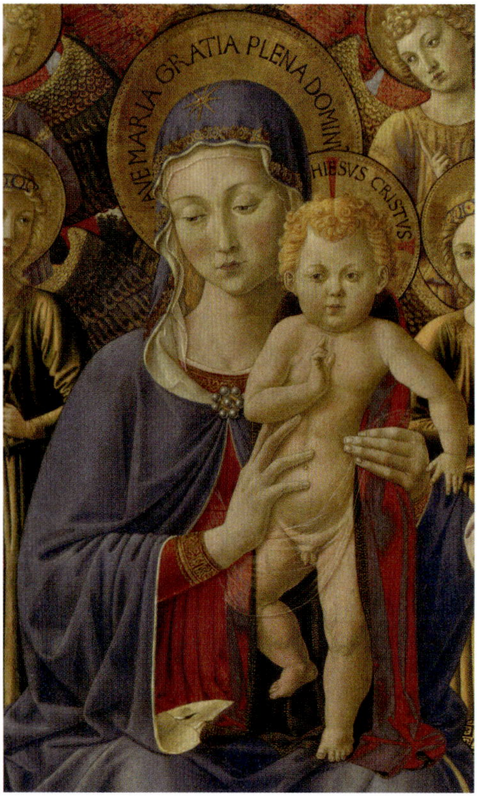

Figure 1 Benozzo Gozzoli, *The Virgin and Child Enthroned among Angels and Saints*, 1461–2, egg tempera on wood, 161.9 × 170.2 cm, © National Gallery, London, NG283. Detail showing darkening of vermilion in the lighter areas of the red drapery around the Christ Child.

this subject carried out at the National Gallery, London, and on examples in paintings dating from before 1800.[1]

The blackening of vermilion

The darkening of the vermilion (red mercury sulphide, α-HgS) in Benozzo Gozzoli's *Virgin and Child Enthroned* (Figure 1) is a typical and quite dramatic example of the alteration of this pigment. As early as the 1st century, Pliny was speculating on the causes, saying that 'cinnabar' (the name now used for the mineral form) was damaged by sunlight, and suggesting that the blackening could be prevented by coating the painting with wax. For a long time the received opinion on the mechanism of this discoloration was that the darkening was simply a light-induced transformation to the black polymorph of mercury sulphide metacinnabar (β-HgS), but several key pieces of work over the last two decades have greatly increased our understanding of the process.

Firstly, some experiments on test samples established that various impurities such as chlorides and other halides, as well as thiosulphates, can accelerate the blackening (Daniels 1987). Vermilion made by the wet process, discovered in 1687, can contain these impurities introduced during manufacture and is therefore inherently unstable, but the dry process vermilion used in earlier paintings can still blacken severely even though it is very pure; it is the latter that is considered here. Another key piece of work was that carried out in the mineralogy field by John McCormack (2000), which showed that the cinnabar in certain mineral deposits that contains some chloride is photosensitive - it blackens when exposed to sunlight. McCormack additionally demonstrated through laboratory experiments that pure vermilion or cinnabar could also react with chloride to form mercury-sulphur-chloride compounds. One specific compound among these, corderoite ($Hg_3S_2Cl_2$), was found to be photosensitive. Further analysis identified calomel (Hg_2Cl_2) as a product after light exposure, together with other compounds that it was not possible to characterise, but which must have been black or grey since the overall bulk colour of the product of these reactions was grey.

Alongside the research on test samples and mineral specimens, examination of paintings in the National Gallery and analyses of samples had noted that areas of deteriorated vermilion were not black (as they had generally been described) but grey. A white or greyish crust always lay over darker black particles, as could be seen in unmounted paint fragments and cross-sections (illustrated in Spring and Grout 2002), explaining why the deteriorated areas often have a purple-grey appearance, the effect of a grey veil over unchanged red beneath. Elemental mapping of cross-sections using energy dispersive X-ray (EDX) analysis in the scanning electron microscope (SEM) revealed that the white material on the surface was mercury chloride, establishing that in paintings, too, chloride is involved in the alteration. Raman analysis further identified calomel (Hg_2Cl_2) among the degradation products in the crust, exactly as found in the mineralogy research by McCormack, suggesting that an analogous process was occurring (Spring and Grout 2002).

The unaltered vermilion below the surface is very pure and does not contain chlorine, indicating that the source must be external pollutants, most probably dirt. Chlorine is certainly one of the components of dirt, as was confirmed by EDX analysis of dirt collected in an air filter from one of the rooms in the National Gallery. Some sodium was detected, so the chlorine was most probably present as sodium chloride. Analysis by gas chromatography-mass spectrometry (GC-MS) detected squalene, a major component of skin grease and therefore an indicator for skin debris; the sodium chloride probably has the same origin (Spring and Grout 2002: 55). EDX analysis of dirt in other museums has given similar results.

From this work it could be proposed that chloride from dirt or other pollutants was reacting with vermilion to form photosensitive mercury sulphochlorides such as corderoite, which were then undergoing photodegradation to form white calomel and a black product (Spring and Grout 2002: 56), but standard laboratory techniques were not able to detect the intermediate products that McCormack had seen, nor was it possible to identify the composition of the black particles. It was only through analysis with synchrotron techniques that it could be established that mercury sulphochloride intermediates, including corderoite, were indeed present. In the samples analysed these compounds appeared yellowish and were located just at the surface of the unaltered vermilion (Radepont *et al.* 2011; Da Pieve *et al.* 2013). Further studies have provided a more detailed and refined understanding of the chemical mechanism and the role of the environment, and have also convincingly shown that the black product is not metacinnabar but metallic mercury (Anaf *et al.* 2013).

The fact that dry process vermilion is not inherently unstable, and that not only light but also an atmospheric pollutant is involved in the deterioration, is also consistent with its seemingly unpredictable behaviour. In Benozzo Gozzoli's *Virgin and Child Enthroned* (Figure 1), for example, only the lightest areas of the red drapery around the Christ Child - painted with vermilion alone - have discoloured, while the deeper more shadowed parts of the modelling where the vermilion was glazed with red lake are well preserved. This glaze may have protected the vermilion from light to some extent, but it could also have acted as a physical barrier to chloride ions from the environment. Similarly, samples from the areas of the modelling that have not discoloured in Saint Jerome's robe in Giovanni dal Ponte's altarpiece, *The Ascension of Saint John the Evangelist* (about 1420-24, National Gallery, London, NG580), were found to have some red lead mixed with vermilion; this too seems to protect the pigment (Spring and Grout 2002: 56-8).

The conservation history of the paintings is also an important factor. The San Pier Maggiore altarpiece (1370-71, National Gallery, London, NG569), by Jacopo di Cione and workshop, was cleaned in the 1980s for the first time since it entered the National Gallery in 1857. Heavy deposits of chloride-containing dirt had collected in the extremely thin surface coating that was present on the painting, as could be seen from photographs taken during cleaning. It seems to have been cleaned in Italy in the 19th century before it was acquired by the National Gallery, when only a thin coating was applied in the belief that early Italian paintings would not have had a varnish. The dirt was therefore in almost direct contact with the paint, and it is probably this that is responsible for the poor state of areas of vermilion in this painting (Spring and Grout 2002: 58).

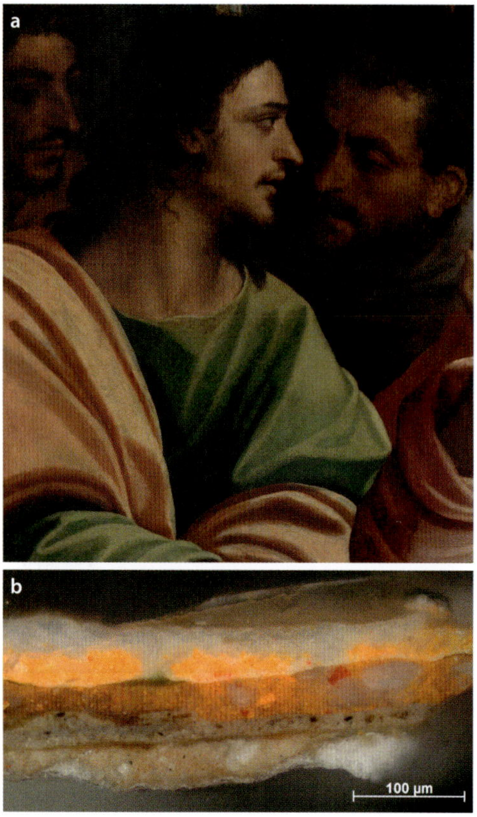

Figure 2 Sebastiano del Piombo, *The Raising of Lazarus*, *c*.1517–19, oil on canvas, transferred from wood, 381 × 289.6 cm, © National Gallery, London, NG1. (a) Detail showing the figure at the left edge of the painting, and (b) a cross-section from the orange drapery in which the upper part of the top orange paint layer is white due to alteration of red lead.

These findings suggest that it is desirable to limit contact between vermilion and dirt, or other sources of chloride ions, an aim that has consequences for both preventive and interventive conservation procedures. At the National Gallery, for example, following any snowfall, 'tac' mats are placed by the entrances in order to limit the amount of salt brought in from outdoors on visitors' shoes. In addition, when cleaning areas of discoloured vermilion it is important to understand that it is not just a discoloration to black – the white material on the painting is also a degradation product of the pigment rather than some other accretion.

The alteration of red lead

Another red pigment that can be unstable is red lead (lead tetroxide), although its deterioration in paintings can be more difficult to recognise and it is also rather more rarely seen. It can either become lighter or darken, depending on the conditions to which it has been subjected, but in easel paintings it is the former that predominates (Saunders *et al.* 2002). The reaction of this pigment with an oil medium to form lead soaps is well known (see, for example, Higgitt *et al.* 2003), but it can also lighten more dramatically than happens with soap formation, as has occurred in the orange drapery of the figure at the left edge of Sebastiano del Piombo's *The Raising of Lazarus* (Figure 2). The lower part of the top paint layer in the cross-section is still red, but the upper part has turned completely white, giving the drapery overall a salmon pink colour. Experiments on test samples show that this lightening occurs at higher levels of humidity, with the red lead turning to basic lead carbonate, the same chemical compound as lead white pigment. The stability can also be influenced by variations in composition resulting from the manufacturing process, since variable amounts of one of the starting materials, lead oxide (PbO), can be present as residues. As with vermilion, analyses with synchrotron techniques have enhanced our understanding of the exact chemical mechanism, clarifying the role of both light and humidity, and demonstrating that lead oxide increases the photosensitivity of red lead (Vanmeert *et al.* 2015), especially in the presence of bicarbonates, which also encourage the photodecomposition and result in the formation of basic lead carbonate (Ayalew *et al.* 2016).

Fading of red lake pigments

The fading of red lake pigments is very familiar, one example among many being in *Henry Dawkins* by Maurice-Quentin de La Tour (pastel on paper mounted on canvas, *c.*1750, National Gallery, London, NG5118), where the difference in colour at the bottom edge where it has been protected from light gives a good indication of the extent to which these pigments can alter.[2] The stability of a red lake depends on the dyestuff used to prepare it, as well as on the recipe, with the substrate having a strong influence. Those most commonly found in paintings are brazilwood, madder, lac, kermes and cochineal. Brazilwood and lac lakes were generally prepared directly from the dyestuff source, but for the others the dyestuff was often extracted from wool or silk textile shearings, another factor affecting the stability (Kirby *et al.* 2006). Artificial ageing experiments on test samples have shown that brazilwood is the least stable and madder is the most lightfast (Saunders and Kirby 1994).

Mixtures of these red lakes with an increasing proportion of lead white were made for these experiments, and the faded samples show an effect, often noted on paintings, which is that as the lake fades the mid-tones become just as pale as the lightest hues (Saunders and Kirby 1994). This can be seen in Zanobi Strozzi's *Annunciation* (Figure 3), where a kermes lake has faded, and only the deepest shadows of the Virgin's dress are still strongly red, while the mid-tones and highlights are now more or less indistinguishable. The dress originally had a mordant-gilded pattern which has protected the red lake from light, and as the gilding has now been mostly lost we can see the well-preserved red left behind in the same pattern (Kirby *et al.* 1997). A more subtle effect seen in these experiments was that the red lakes did not simply lose their colour: the brazilwood moved as it faded towards a yellowish hue, and while the lakes prepared from insect dyestuffs also initially moved towards yellow, after further fading they became rather greyish and slightly purple in tone. A possible example of this in a painting can be found in Jan van Huysum's *Flowers in a Terracotta Vase* (1736-7, oil on canvas, National Gallery, London, NG796), in the mid-tones of the tulips at the top right of the spray of flowers.

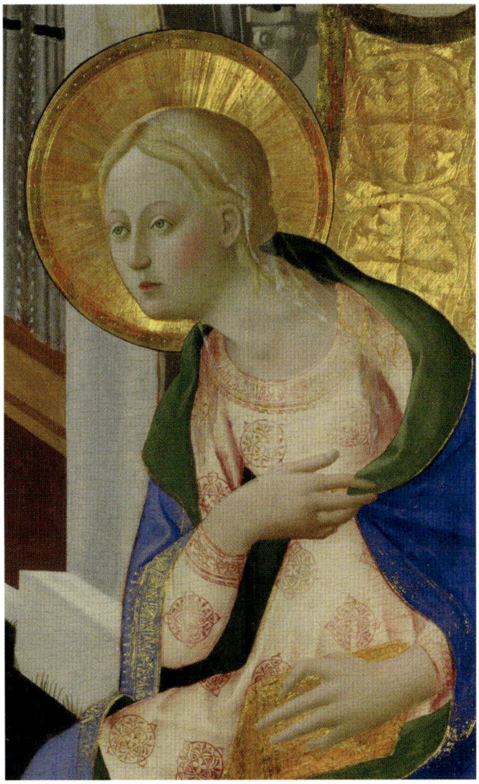

Figure 3 Zanobi Strozzi, *The Annunciation*, *c.*1440–45, egg tempera on wood, 104.5 × 142 cm, © National Gallery, London, NG1406. Detail showing the faded pink dress of the Virgin.

Red lake was very often a component of purple mixtures, and similar ageing experiments on test samples to those carried out on red lakes alone demonstrated how the fading of the red component can shift the hue from a relatively warm purple shade to a very cool lilac that is almost blue. The lilac drapery of the nymph at the left of Titian's *Diana and Actaeon* (Figure 4) is painted in mixtures of lead white, red lake and ultramarine, and a cross-section shows more purple paint immediately beneath ultramarine blue at the surface. This could perhaps be the same effect seen in the test samples, and the lilac drapery might at first have been a warmer purple, but it also indicates the difficulty of translating what is observed in experiments to what is seen in paintings, as it is known from other works by Titian that

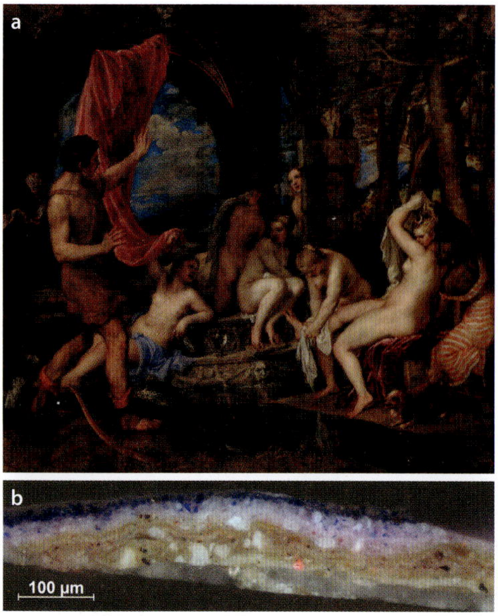

Figure 4 (a) Titian, *Diana and Actaeon*, 1556–9, oil on canvas, 184.5 × 202.2 cm, © National Gallery, London, NG6611. (b) A cross-section from the lilac drapery of the nymph at the left of the painting.

he did sometimes deliberately underpaint his blues with a pink or purple tone. It is not therefore possible to be certain that the rather blue appearance is due to fading of a red lake (Dunkerton and Spring 2015).

Verdigris-containing oil paints

Some of the greens in Titian's *Diana and Actaeon* (Figure 4) also seem to have altered – in the trees at the top right, for example, where the darker tones painted with verdigris have become brown (the lightest leaves, which contain malachite are still bright green). A cross-section prepared from a sample showed that the discoloration is only at the very surface, with verdigris that is still green visible beneath (Dunkerton and Spring 2015: 26). While this browning can be a change in the paint itself, it can also be a consequence of reaction of an early varnish (either oil-resin or resin alone) with copper in the pigment forming a thin deep brown zone at the interface with the paint. Analysis of cross-sections with attenuated total reflectance-Fourier transform infrared (ATR-FTIR) microspectroscopic imaging has proved useful for investigating these changes. In Jan van Eyck's *Portrait of Giovanni(?) Arnolfini and his Wife* (1434, oil on oak, National Gallery, London, NG186), for example, although the green of the woman's dress seems well preserved, a cross-section showed a very thin brown layer at the surface of the verdigris-containing paint. In the past it was assumed that in Early Netherlandish paintings, translucent green glazes such as this were 'copper resinate' made by heating a copper salt with an oil-resin mixture, but the current prevailing opinion is that they are in fact composed of verdigris. This was confirmed for this painting by the spectra from ATR-FTIR imaging, which showed evidence of unreacted verdigris pigment as well as a clear indication that it has reacted with the oil to form copper soaps. In this case a little resin was detected by GC-MS in addition to oil, but ATR-FTIR imaging was able to show that although some 'copper resinate' was present, it was located only at the very surface of the cross-section and so must have been formed from reaction of the paint with an old resinous varnish that had been present at some time on the surface (Spring and Morrison forthcoming).

The state of preservation of verdigris glazes can be somewhat variable, but where they have discoloured badly they have often been applied on top of a light paint, as in Jacopo Bassano's *The Good Samaritan*, where the stripes on the scarf are now brown (Figure 5). A cross-section shows the original green colour still present beneath, but hidden by the deterioration at the surface. ATR-FTIR imaging again showed that the paint contains verdigris that has reacted with the oil binder to form copper carboxylates, but in this case there was no indication of any 'copper resinate' at the surface, so the browning is not simply reaction with an old varnish but a more fundamental change in the paint (Dunkerton and Spring 2014). The chemical changes associated with this degradation are still not well understood, although the most recent hypotheses, based on analyses of test samples with electron paramagnetic resonance and X-ray absorption spectroscopy as well as other techniques, propose that the colour change may be linked to the formation of copper(I) in the copper complexes of the pigment-oil system (Santoro *et al.* 2014).

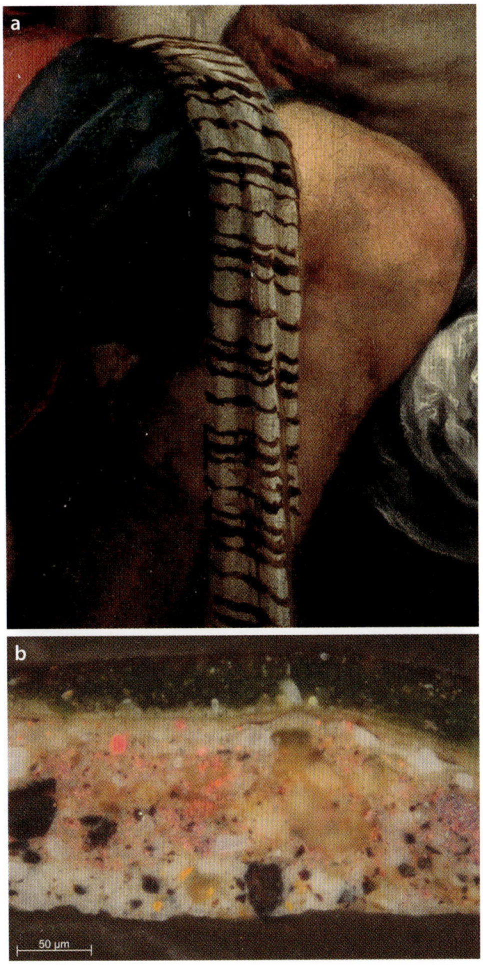

Figure 5 Jacopo Bassano, *The Good Samaritan*, *c.*1562–3, oil on canvas, 102.1 × 79.7 cm, © National Gallery, London, NG277. (a) Detail showing the browned green stripes on the Samaritan's scarf, and (b) a cross-section from the same area.

Blanching of paint films: yellow lake and ultramarine

A common way to achieve green was by combining blue and yellow, and in the 17th and 18th centuries in particular these mixtures often included yellow lake pigments, which tended to be especially fugitive because they were generally prepared with a substrate consisting predominantly of chalk, rather than the alumina found for red lakes (Kirby and Saunders 1998). The effects of this are most obvious where foliage has become blue as the yellow has faded, as can be seen in Adam Pynacker's *Landscape with Sportsmen and Game* (*c.*1665, oil on canvas, Dulwich Picture Gallery, DPG086) in the large leaves in the centre of the foreground, and also in the leaves in Jan van Huysum's *Flowers in a Terracotta Vase* (1736–7, oil on canvas, National Gallery, London, NG796). The deterioration of yellow lake in an oil binder often involves not only fading but also some breakdown of the paint film causing blanching, as has happened in many paintings by Aelbert Cuyp, including '*The Large Dort*' (Figure 6). In a cross-section from the deep olive green foreground landscape, a mixture of yellow lake, vivianite and earth pigments, the paint is lighter at the surface where the yellow lake component has lost its colour, but the paint has also broken up and is scattering light (Spring and Keith 2009).

The alteration of paint containing ultramarine is also generally manifested as blanching. In Jan van Eyck's *Portrait of Giovanni(?) Arnolfini and his Wife*, the damask pattern in the woman's blue sleeves is now barely visible because the final thin layer of ultramarine has deteriorated, becoming patchy and lighter than it was originally so that the darkest blues appear very similar to the mid-tones.[3] Deterioration of ultramarine is often stated to be loss of colour in the blue mineral component lazurite under acidic conditions, sometimes called 'ultramarine sickness', but in fact this degradation is still not fully understood. Whatever changes might be occurring in the

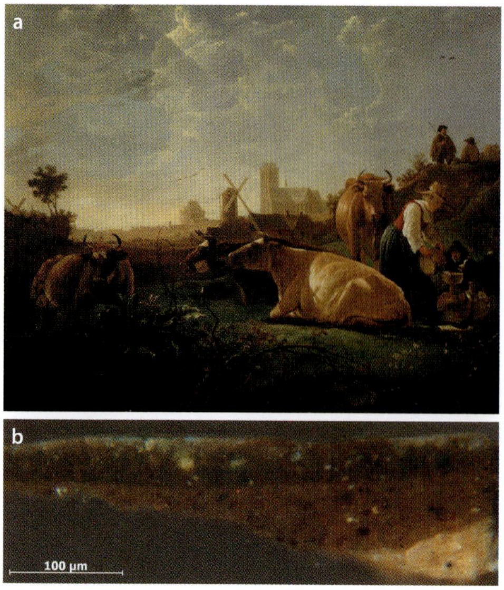

Figure 6 (a) Aelbert Cuyp, *A Distant View of Dordrecht, with a Milkmaid and Four Cows, and Other Figures ('The Large Dort')*, c.1650, oil on canvas, 157.5 × 197 cm, © National Gallery, London, NG961. (b) A cross-section from the olive-green foreground landscape at the left.

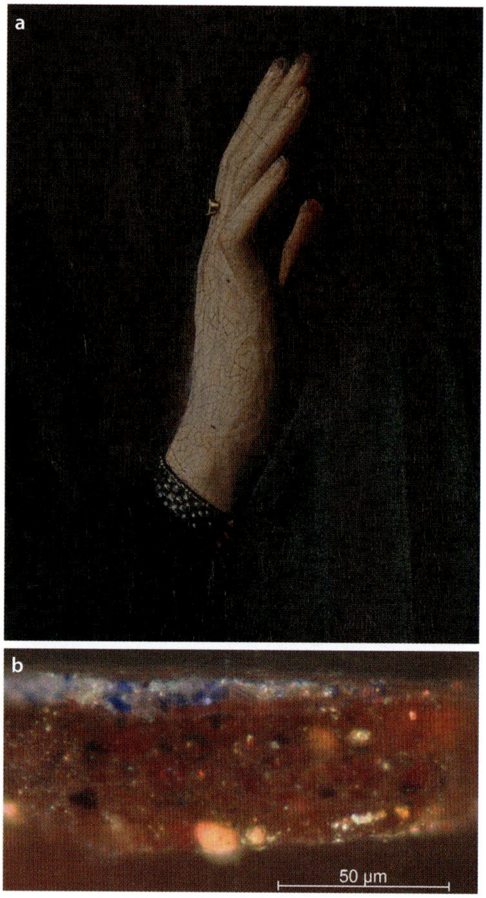

Figure 7 Jan van Eyck, *Portrait of Giovanni(?) Arnolfini and his Wife*, 1434, oil on oak, 82.2 × 60 cm, © National Gallery, London, NG186. (a) Detail showing Giovanni Arnolfini's dark purple robe, with blanching in the shadows, and (b) a cross-section from the same area.

pigment, examination of paint samples from areas of deteriorated ultramarine in paintings shows, in addition, evidence of breakdown of the binding medium, and other observations from both samples and paintings help to give some idea of the factors that might be involved. Paintings by certain artists seem to suffer more often from this phenomenon than those by other artists, suggesting that an individual's technique might be important. Jan van Eyck, for example, seems to have used more finely ground ultramarine than some of his contemporaries, including Rogier van der Weyden. This would have given his paint better handling properties, as coarse ultramarine gives a paint that is not very manageable and difficult to apply, but this might be a factor in the poor state of the blues in quite a number of his works (Spring and Morrison forthcoming).

This deterioration is most noticeable in Giovanni Arnolfini's purple robe (Figure 7). The blanching of the ultramarine paint makes the half shadows appear lighter than they would have been originally, reversing the tonal values of the modelling. A cross-section shows a deep plum-coloured paint containing large red lake particles mixed with only a little ultramarine, on top of which is the very thin layer that forms the half shadow of the robe. This final glaze is pale in appearance, but since there is no lead white mixed with the ultramarine, this is caused by the scattering of light in this broken up and deteriorated paint. Another observation was that calcium oxalate was detected by ATR-FTIR imaging in this layer (Spring and Morrison forthcoming), as has been found regularly in ultramarine-containing paint

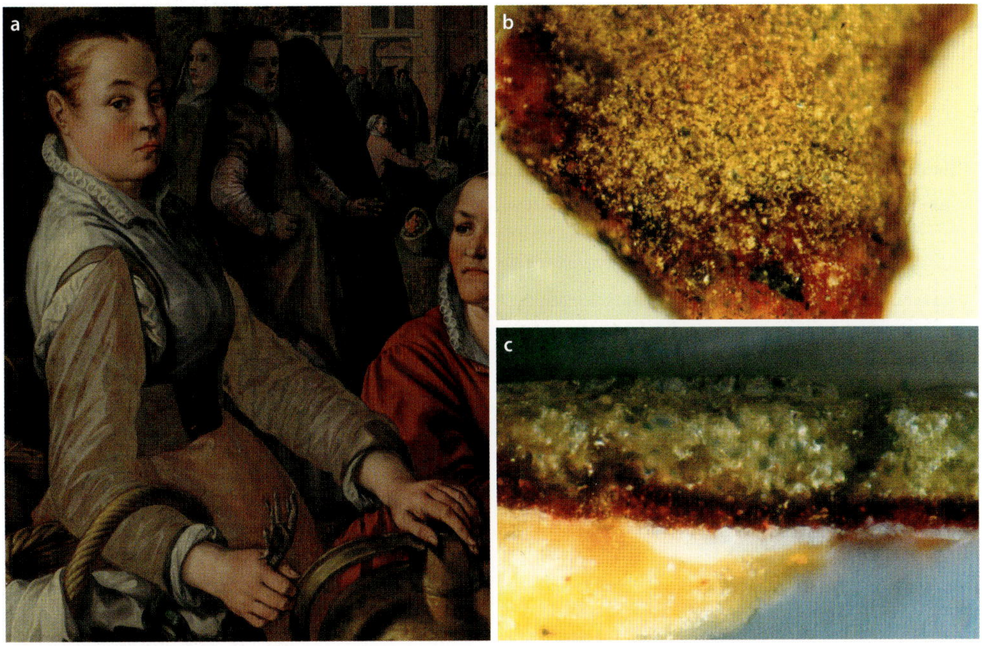

Figure 8 Joachim Beuckelaer, *The Four Elements: Air*, 1570, oil on canvas, 158 × 216 cm, © National Gallery, London, NG6587. (a) Detail showing the woman at the left, with sleeves painted with smalt (now deteriorated) over red lake; (b) unmounted sample from the sleeve; (c) a cross-section of the same sample.

when analysed by FTIR (Higgitt and White 2005). This is not likely to have been an original component of the paint, and as oxalates can be an indication of extreme degradation of the medium, this may prove to be significant in understanding the deterioration process.

Deterioration of smalt

Blanching of the paint (as well as other defects) also often accompanies the loss of colour from the blue cobalt glass pigment smalt when it deteriorates, as has happened in the series of panels by Joachim Beuckelaer, *The Four Elements*, for example in the sleeves of the woman at the left of *Air* (Figure 8). This is particularly evident in an unmounted paint sample from this area and in the cross-section – not only does the paint layer appear very broken up but there is also a crust on the surface adding to the blanched appearance. It is clear that it is not only the pigment that is affected by the deterioration. It is well known that when smalt-containing oil paint deteriorates, the alkaline component of the glass, which in the case of smalt is potassium, is leached from the pigment and then reacts with the binding medium to form potassium soaps. These soaps can then migrate up to the surface, forming a crust (Spring *et al.* 2005; van Loon 2008).

The extensive research that has been carried out on the deterioration of window and vessel glass is useful when considering smalt. This has found that the leached alkali goes on to react with anions in the environment to form various salts such as arcanite (potassium sulphate), syngenite (potassium calcium sulphate), as well as calcium carbonate, sulphate and oxalate. The source of calcium in these complex crusts is not the glass itself but airborne particles in dirt that are deposited on the surface and become incorporated into the crust through reaction with the somewhat alkaline and hygroscopic salts initially formed. Biodeterioration is also thought to play a role, manifested by the presence of calcium oxalate. Analyses of the crust on the surface of degraded smalt-containing paint have revealed similar compounds, including calcium oxalate and sometimes potassium lead sulphate (Spring *et al.* 2005; van Loon *et al.* 2012). The potassium soaps formed initially on reaction with fatty acids in the oil are water soluble, which could have some implications for the use of aqueous cleaning agents during conservation treatment, but if a crust has formed at the surface this may not be a significant risk, since it is likely to protect the paint beneath at least to some extent because it is very insoluble.

The relationship between colour and potassium content is clear when comparing EDX spectra from well-preserved particles and those that have become pale, with the latter being noticeably depleted in potassium. The potassium content can therefore be measured to distinguish between smalt that is degraded and that which is a pale variety or grade, and to gain some idea of the threshold at which the pigment loses its colour. Well-preserved smalt usually contains between 12 and 15 weight % potassium oxide. Each cobalt ion needs two potassium ions to maintain the blue colour (Robinet *et al.* 2011). From quantitative SEM-EDX analysis on a cross-section from the sky in Rembrandt's *Anna and the Blind Tobit* (*c.*1630, oil on oak, National Gallery, London, NG 4189) this could be estimated to equate to around 7 weight % potassium oxide, as measured in a particle which is pale but still has a little colour, and as a rule of thumb smalt with less than 5% potassium oxide can be considered to be definitely degraded (Spring forthcoming).

Cobalt is not generally lost from the pigment on degradation, so quantitative measurement of the cobalt content gives some idea of the original intensity of colour (Spring *et al.* 2012), although particle size does have an influence so also needs to be considered. In Rembrandt's *Anna and the Blind Tobit*, the smalt contains around 5 weight % cobalt oxide and is therefore a relatively high grade. Similar analyses of the smalt in the underpaint for the sky in Titian's *Diana and Actaeon* (Figure 4) show that it too would have had a strong blue colour initially, as it contains 4–5 weight % cobalt oxide, but it is certainly degraded since it contains only about 2 weight % potassium oxide. This underpaint was left exposed in some places, for example behind the trees in the upper right of the painting, and now that it has become grey it contrasts much more strongly with the other areas of the sky where it has been covered with a layer of ultramarine (Dunkerton and Spring 2015).

Smalt was used extensively in Dutch 17th-century paintings and *The Astronomer* by Ferdinand Bol is no exception (Figure 9); there is smalt in the paint mixtures used for the tablecloth, curtain and globes, all now degraded. In the tablecloth, analysis of a sample found a mixture of smalt with large amounts of yellow lake and a little yellow earth, which must originally have given a green colour of some kind. Thanks to a collaboration with Delft University of Technology it was possible to examine this painting with a Bruker M6 macro X-ray fluorescence scanner, producing maps of chemical elements that give far more information on how these pigments are distributed and what has been lost in terms of modelling, for example in the now rather dull mustard brown tablecloth. The cobalt map reveals the location of smalt, present in most parts of the tablecloth, and the calcium map locates the yellow lake, as well as any bone black. The iron XRF map shows that an earth pigment seems to be mainly in the lightest areas, and the copper map reveals small strokes that are perhaps a pattern that is now hardly visible. The cobalt and calcium maps in particular also indicate the extent of colour change in the curtain, showing that the modelling would have been rather more evident, with more detailed folds and creases (Figure 9). It is clear that macro XRF scanning is proving a useful new tool for investigating colour changes in paintings in a more comprehensive way across a painting than is possible with a few samples, as has also been demonstrated for several paintings by Rembrandt that have been examined with this technique and that also exhibit deterioration of smalt (Janssens *et al.* 2016).

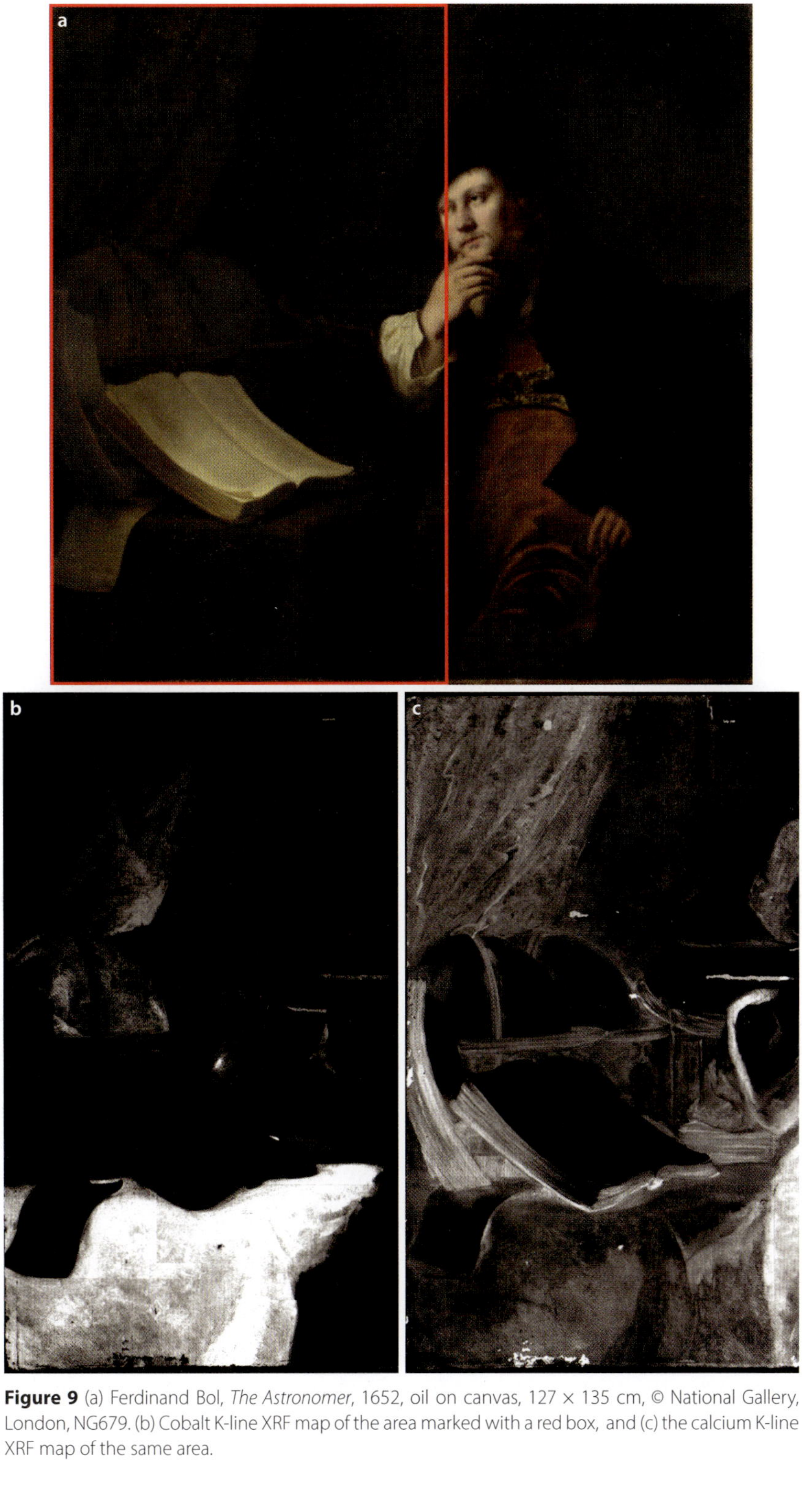

Figure 9 (a) Ferdinand Bol, *The Astronomer*, 1652, oil on canvas, 127 × 135 cm, © National Gallery, London, NG679. (b) Cobalt K-line XRF map of the area marked with a red box, and (c) the calcium K-line XRF map of the same area.

Conclusions

Research carried out using the sophisticated analytical techniques now available – those at synchrotron facilities in particular – are gradually helping us to better understand the processes involved in pigment deterioration, although there is much still to discover. The examples mentioned here demonstrate how ATR-FTIR imaging can help to probe the status of browning in verdigris glazes, as well as the value of quantitative SEM-EDX analyses of smalt, in determining whether the pigment has degraded or is instead a low grade that was always pale (from the potassium content) – a question that helps to define the artist's intention and to estimate, from the cobalt content, the initial intensity of colour. New analytical imaging techniques such as macro XRF scanning provide other information, in a non-invasive manner on the painting itself, that also helps us to consider how the painting appeared before deterioration of certain pigments. This improved knowledge of the changes that have occurred over time on a painting can help to inform conservation treatment, while the observations from samples refine understanding of what is seen on the surface in deteriorated areas of paint, such as the grey crust formed on vermilion, and the blanching seen with degraded ultramarine, red lake and smalt. The detailed understanding of chemical mechanisms, meanwhile, clarifies the role of light, humidity and pollutants in initiating or exacerbating the deterioration, and can inform future preventive conservation strategies.

Acknowledgements

I am grateful to Delft University of Technology for collaboration on macro XRF scanning of the painting by Ferdinand Bol, especially Joris Dik and Annelies van Loon, and also to colleagues at the National Gallery, Nelly von Aderkas (for processing of the XRF images) and Rachel Morrison (for ATR-FTIR imaging of the samples from the paintings by Jacopo Bassano and Jan van Eyck).

Notes

1. For more modern materials see the papers by Ormsby and Lee, and Hendriks, in this vol.
2. See www.nationalgallery.org.uk/paintings/maurice-quentin-de-la-tour-henry-dawkins (accessed 11 March 2016).
3. See www.nationalgallery.org.uk/paintings/jan-van-eyck-the-arnolfini-portrait (accessed 11 March 2016).

References

Anaf, W., Janssens, K. and De Wael, K. (2013) 'Formation of metallic mercury during photo-degradation/photodarkening of α-HgS: electro chemical evidence', *Angewandte Chemie – International Edition* 52: 12568–71.

Ayalew, E., Janssens K. and De Wael, K. (2016) 'Unravelling the reactivity of minium towards bicarbonate and the role of lead oxides therein', *Analytical Chemistry*: DOI:10.1021/acs.analchem.5b02503

Da Pieve, F., Hogan, C., Lamoen, D., Verbeeck, J., Vanmeert, F., Radepont, M., Cotte, M., Janssens, K., Gonze, X. and Van Tendeloo, G. (2013) 'Casting light on the darkening of colors in historical paintings', *Physical Review Letters* 111(20): 208302.

Daniels, V. (1987) 'The blackening of vermilion by light', in *Recent Advances in the Conservation and Analysis of Artifacts, Jubilee Conference Papers*. London: Institute of Archaeology, University of London, 280–82.

Dunkerton, J. and Spring, M. (2014) 'Developments in the technique of Jacopo Bassano's works in the National Gallery, London', in *Jacopo Bassano, i figli, la scuola, l'eredità, Atti del Convegno Internazionale di Studio, Bassano del Grappa, Museo Civico – Padova, Università degli Studi, Archivio Antico del Bo, 30 marzo–2 aprile 2011, Bollettino del Museo Civico* 30–34 (2009–2013), vol. III, 752–65, 901–7.

Dunkerton, J. and Spring, M., with contributions from Billinge, R., Howard, H., Macaro, G., Morrison, R., Peggie, D., Roy, A., Stevenson, L.

and Von Aderkas, N. (2015) 'Titian after 1540: technique and style in his later works', *National Gallery Technical Bulletin* 36: 6-39.

Higgitt, C. and White, R. (2005) 'Analyses of paint media: new studies of Italian paintings of the fifteenth and sixteenth centuries', *National Gallery Technical Bulletin* 26: 88–97.

Higgitt, C., Spring, M. and Saunders, D. (2003) 'Pigment-medium interactions in oil paint films containing red lead or lead-tin yellow', *National Gallery Technical Bulletin* 24: 75–95.

Janssens, K., van der Snickt, G., Alfeld, M., Noble, P., van Loon, A., Delaney, J., Conover, D., Zeibel, J. and Dik, J. 2016. 'Rembrandt's *Saul and David* (c. 1652): use of multiple types of smalt evidenced by means of non-destructive imaging', *Microchemical Journal* 126: 515–23.

Kirby, J. and Saunders, D. (1998) 'Sixteenth- to eighteenth-century green colours in landscape and flower paintings: composition and deterioration', in A. Roy and P. Smith (eds), *Painting Techniques. History, Materials and Studio Practice, Contributions to the Dublin IIC Congress, 7-11 September 1998*. London: International Institute for Conservation, 155–60.

Kirby, J., Saunders, D. and Cupitt, J. (1997) 'Colorants and colour change', in T. Bakkenist, R. Hoppenbrouwers and H. Dubois (eds), *Early Italian Paintings: Techniques and Analysis*. Maastricht: Limburg Conservation Institute, 65–72.

Kirby, J., Saunders, D. and Spring, M. (2006) 'Proscribed pigments in northern European Renaissance paintings and the case of Paris red', in D. Saunders, J.H. Townsend and S. Woodcock (eds), *The Object in Context: Crossing Conservation Boundaries. Contributions to the Munich IIC Congress, 28 August–1 September 2006*. London: International Institute for Conservation, 236–43.

McCormack, J.K. (2000) 'The darkening of cinnabar in sunlight', *Mineralium Deposita* 35: 796–8.

Radepont, M., de Nolf, W., Janssens, K., Van der Snickt, G., Coquinot, Y., Klaassen L. and Cotte, M. (2011) 'The use of microscopic x-ray diffraction for the study of HgS and its degradation products corderoite (α-$Hg_3S_2Cl_2$), kenhsuite (γ-$Hg_3S_2Cl_2$) and calomel (Hg_2Cl_2) in historical paintings', *Journal of Analytical Atomic Spectrometry* 26(5): 959-68.

Robinet, L., Spring, M. and Pagès-Camagna, S. (2011) 'Investigation of the loss of colour in smalt on degradation in paintings using multiple spectroscopic analytical techniques', in J. Bridgland (ed.), *Preprints of the ICOM-CC 16th Triennial Conference, Lisbon 19-23 September 2011*, paper 1614.

Santoro, C., Zarkout, K., Le Hô, A-S., Mirambet, F., Gourier, D., Binet, L., Pagès-Camagna, S., Reguer, S., Mirabaud, S., Le Du, Y., Griesmar, P., Lubin-Germain, N. and Menu, M. (2014) 'New highlights on the degradation process of verdigris from easel paintings', *Applied Physics A* 114: 637–45.

Saunders, D. and Kirby, J. (1994) 'Light-induced colour changes in red and yellow lake pigments', *National Gallery Technical Bulletin* 15: 79–97.

Saunders, D., Spring, M. and Higgitt, C. (2002) 'Colour change in red lead-containing paint films', in R. Vontobel (ed.), *Preprints of the ICOM-CC 13th Triennial Meeting*. London: James and James, 455–63.

Spring, M. (forthcoming) 'Pale grey or deep blue? Evaluating smalt in Rembrandt's paintings', in *Rembrandt Now: Technical Practice, Conservation and Research, Contributions to the Conference at the National Gallery, London, 13–15 November 2014*. London: Archetype Publications.

Spring, M. and Grout R. (2002) 'The blackening of vermilion: an analytical study of the process in paintings', *National Gallery Technical Bulletin* 23: 50–61.

Spring, M. and Keith, L. (2009) 'Aelbert Cuyp's *Large Dort*: colour change and conservation', *National Gallery Technical Bulletin* 30: 71-85.

Spring, M. and Morrison, R. (forthcoming) 'Jan van Eyck's technique and materials: historical perspectives and contemporary context', in *Proceedings of the Van Eyck Studies Colloquium, Brussels 19–21 September 2012*.

Spring, M., Higgitt, C. and Saunders, D. (2005) 'Investigation of pigment–medium interaction processes in oil paint containing degraded smalt', *National Gallery Technical Bulletin* 26: 56–70.

Spring, M., Kugler, V. and Bean, S. (2012) 'Quantitative energy dispersive X-ray analysis of the blue pigment smalt in the variable pressure scanning electron microscope', in N. Meeks, C. Cartwright, A. Meek and A. Mongiatti (eds), *Historical Technology, Materials and Conservation, SEM and Microanalysis*. London: Archetype Publications in association with the British Museum, 114–22.

van Eikema-Hommes, M. (2004) *Changing Pictures. Discoloration in 15th-17th-century Oil Paintings.* London: Archetype Publications.

van Loon, A. (2008). *Color Changes and Chemical Reactivity in Seventeenth-century Oil Paintings: Molart Reports*. London, Archetype Publications.

van Loon, A., Noble, P. and Boon, J.J. (2012) 'The formation of complex crusts in oil paints containing lead white and smalt: dissolution, depletion, diffusion, deposition', in N. Meeks, C. Cartwright, A. Meek and A. Mongiatti (eds), *Historical Technology, Materials and Conservation: SEM and Microanalysis*. London: Archetype Publications in association with the British Museum, 205–7.

Vanmeert, F., Van der Snickt, G. and Janssens, K. (2015) 'Plumbonacrite identified by X-ray powder diffraction tomography as a missing link during degradation of red lead in a Van Gogh painting', *Angewandte Chemie - International Edition* 54: 3607-10.

Author's address

Marika Spring, Scientific Department, National Gallery, London, UK (marika.spring@ng-london.org.uk)

Colour change and restoration

Larry Keith

Introduction

Most Old Master paintings have undergone significant changes in colour, often in ways that are localised within the whole of a given composition. Scientific research on the causes and mechanisms of such changes continues to provide important discoveries concerning these alterations, but rarely gives an authoritative calculation of the degree of the shifts that have occurred and, importantly, their aesthetic significance. In treating pictures, the conservator must be as informed as possible about these colour changes, but must deploy that knowledge within the wider interpretive context that is a central element of restoration practice. That interpretive function is itself a complex mixture of wider cultural attitudes – the so-called 'period eye' – and the more narrowly defined taste of those responsible for particular restorations. For the most part, pigment alterations are accepted within Old Master paintings as an inevitable consequence of ageing, so considerable efforts are made to ensure that their display environment maximises their preservation. However, there are instances when some restoration intervention may also occur, particularly where pigment alteration is disruptive of tonal values or spatial representation more than simply colour relationships. The degree to which we might consider such interventions varies from painting to painting, period to period, and is also affected to some extent by other considerations of condition and context of display.

In the course of study, conservation students are made aware of the changes that can occur to pigments. One of the best-known examples from a British collection is found in the blue leaves of the Dulwich Picture Gallery's *Landscape with Sportsmen and Game* by Adam Pynacker from *c.*1665 (Figure 1). As part of the conservation training at the Hamilton Kerr Institute in Cambridge, copies are made of Old Master paintings that are undergoing treatment: these are not intended to mimic the current state but to recreate the original appearance. There have been some interesting and notable examples of others copying works along such lines, including some measure of replicating technique rather than a simple imitation of present appearance. For example, the fascinating copies made in the 1930s after Renaissance paintings by the Russian painter Nicholas Lochoff, which have been reproduced in Paul Taylor's recent book on understanding changes in the condition of Old Master paintings (Taylor 2015: 6, 8, 20–22). The essentially open-air Imperial Palace in Kyoto displays impressive modern reconstructions, faithful to original techniques and materials, of largely 16th- and 17th-century paintings first made for the spaces; the originals are displayed in an adjoining, climate-controlled museum. Such efforts demonstrate an awareness of these changes, even if imperfectly understood, and their importance for our understanding of Old Master paintings.

Understanding the process and effects of individual examples of colour change, however

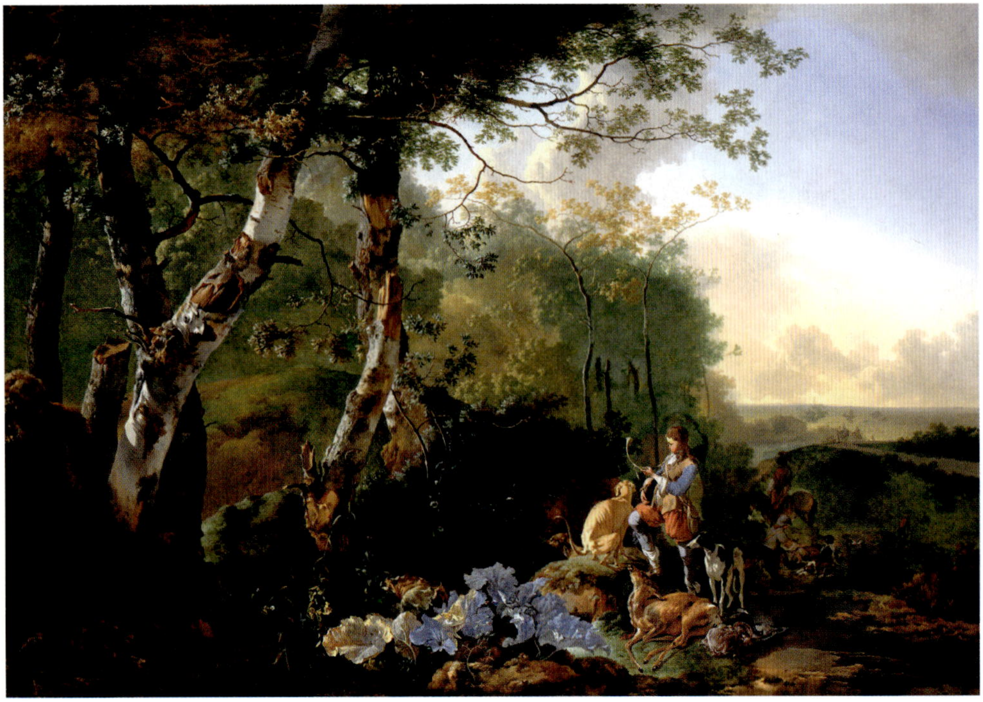

Figure 1 Adam Pynacker, *Landscape with Sportsmen and Game*, *c.*1665, oil on canvas, 137.8 × 198.7 cm, DPG86. (By permission of the Trustees of Dulwich Picture Gallery, London)

interesting, is but the first step in thinking about the wider implications as to how they can affect the comprehension and reading of pictures as a whole. The 1988 exhibition of Sienese painting at the Metropolitan Museum of Art in New York, in which predella panels and miniatures by Liberale da Verona were more or less juxtaposed, provided a powerful lesson on the degree and extent of change which had occurred in the predella panels (Christiansen *et al.* 1988). Of course the intent of painting miniatures and altars is quite different (for example, viewing distance, the kind of viewer scrutiny etc.), but both categories of images had other broader aims in common, including their ways of balancing and distributing colour across the composition. In miniature painting one can often still see the way that the natural stridency of (largely unchanged) vermilion could be reined in somewhat by an unaltered purple-red lake, or the sophistication of the juxtaposing of the inherently different hues imparted by ultramarine and azurite, or the function of greens of greater intensity than are generally seen on panel paintings in their aged state. There are similar comparisons to be made with other Renaissance painters who were accomplished illuminators, such as Fra Angelico and Lorenzo Monaco, while an exhibition held about 10 years later at the National Gallery, London, provided similar opportunities with works associated with Pisanello (Syson and Gordon 2001). These comparisons all emphatically reinforce the same basic lesson: most such panel paintings have changed substantially and irrevocably in their colours and colour relationships, so the practice of cleaning and restoration of Old Master paintings in particular must take place within that understanding. This is not exactly breaking news, but it stands in resolute opposition to any facile notion of restoration making a painting look 'like new', attractive as that idea may be, and so this understanding must

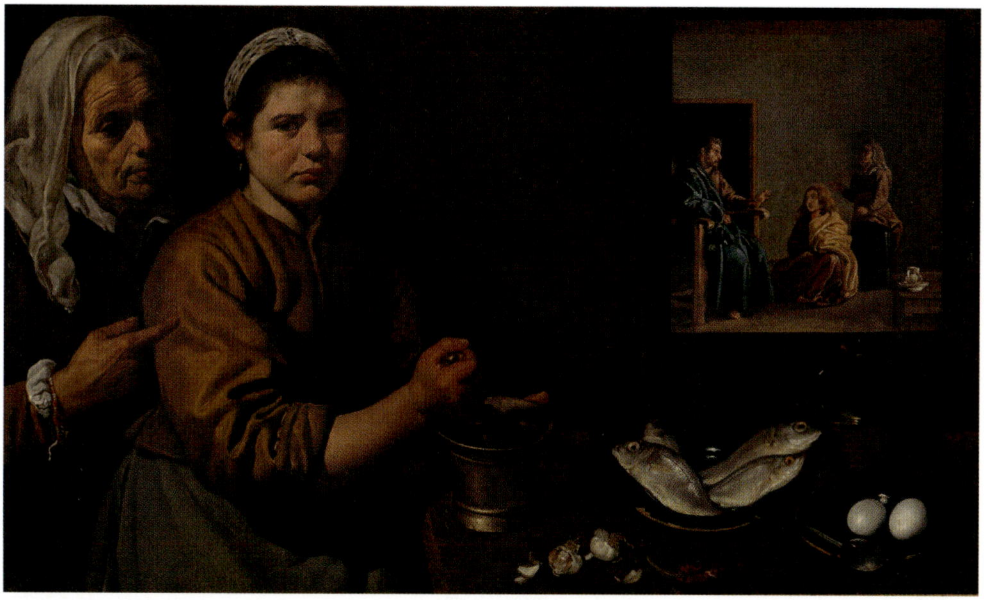

Figure 2 Diego Velázquez, *Christ in the House of Martha and Mary*, *c.*1618, oil on canvas, 60 × 103.5 cm, National Gallery, London, NG1375. After 2011–12 restoration. (© The National Gallery, London)

be the basic foundation on which any ideas on retouching must rest. There must be an understanding and acceptance of a significant amount of change – whether fading, darkening, blanching or increased transparency – in addition to any consideration of damage or loss. Helpfully, recent works such as Margriet van Eikema Hommes' *Changing Pictures* and Paul Taylor's *Condition: The Ageing of Art* have done much to make the nature of these sorts of changes clearer to a far wider audience (Taylor 2015; van Eikema Hommes 2004).

One lesson that may be drawn from looking at such early Italian painter/miniaturists is to keep a mental inventory of easel paintings of apparently exceptional preservation, particularly from later periods where any useful overlap with well-preserved analogous works in other media, such as illuminations, more or less disappears. From the National Gallery, one can contemplate Carlo Dolci's *Adoration of the Kings*, Guido Reni's *Coronation of the Virgin*, or Rubens' *Samson and Delilah* in something like their original splendour.[1] Most conservators probably keep a mental storeroom of similar works from all over the world which can function as beautifully preserved benchmarks to inform thinking on the less fortunate majority of paintings.

For the most part, as conservators or restorers encounter these colour changes, informed by increasing understanding, they sigh and move on. In general they are changes to be managed rather than corrected. Our primary professional responsibility is to be as knowledgeable as possible on the mechanisms of change, making informed judgements about environmental control and treatment choices to reduce the likelihood of their being exacerbated. This, of course, is an area in which the ongoing research undertaken by our scientific colleagues is essential and invaluable. There is also a whole other discussion to be had about the balance between tonal contrasts and colour relationships in the context of removing or reducing discoloured coatings from paintings which have changed in different ways across their surfaces, but this paper is more concerned with the actual restoration of works, their retouching and not their cleaning.

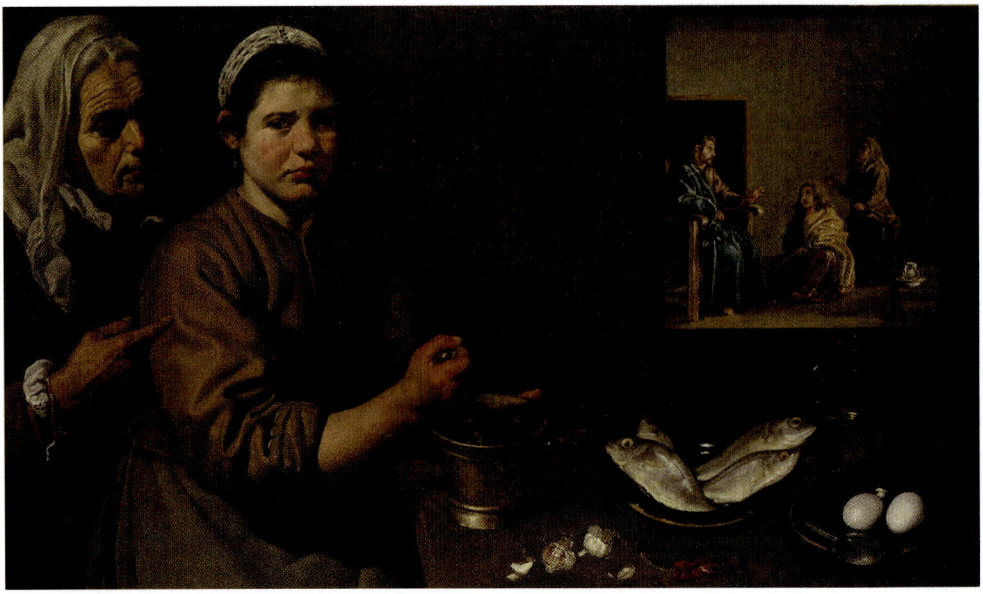

Figure 3 Diego Velázquez *Christ in the House of Martha and Mary*, NG1375, *c*.1618: before 2011–12 restoration. (© The National Gallery, London)

A fairly recent restoration undertaken at the National Gallery on *Christ in the House of Martha and Mary* by Diego Velázquez (Figure 2) is a good treatment-based example. The project raises some interesting issues concerning the application of knowledge of a painter's technique and materials for the resolution of some fundamental problems. In this case it not only explains some perplexing features of the painting's surface, but also provides the intellectual underpinning for the basic interpretive aspects of the restoration itself.

The painting entered the collection in 1892, and was most recently treated in 2011/12. Gallery curators, scientists and conservators looked closely at all works by Velázquez from the collection as part of the preparation for the National Gallery's 2008 Velázquez exhibition (Carr *et al.* 2006),[2] and at the time it was noted how badly the retouching had discoloured on this painting. Close examination suggested that the retouching was also lacking precision and clarity, partly from a change in the materials used, but also from what seemed to be a misunderstanding of the painter's materials and intentions (Figure 3).

Pacheco's 1649 *Arte de la pintura* describes the ground found on all the paintings at the National Gallery from Veláquez's time in Seville: the so-called *tierra de sevilla*, a clay-based earth, rich in silicates, onto which the thick and confident application of paint so characteristic of his early years was executed (Keith 2006; Veliz 1996; Garrido Perez 1992). The ground is often left exposed, although perhaps not as extensively or systematically as is the case in the Caravaggesque painting to which it is often so loosely compared. However, in most cases when it is left visible we see that it has become chalky and pale light grey, a colour and tone which makes little sense in the context of the surrounding paint. The reasons for this are unclear, although superficially the changes slightly resemble the blanching of green earth pigments which is often seen in 17th-century paintings made in Rome, and the phenomenon may be related to the moisture and heat employed in historic glue-paste linings. Similar passages are readily visible on the National Gallery's other early Velázquez paintings – the *Immaculate Conception* and

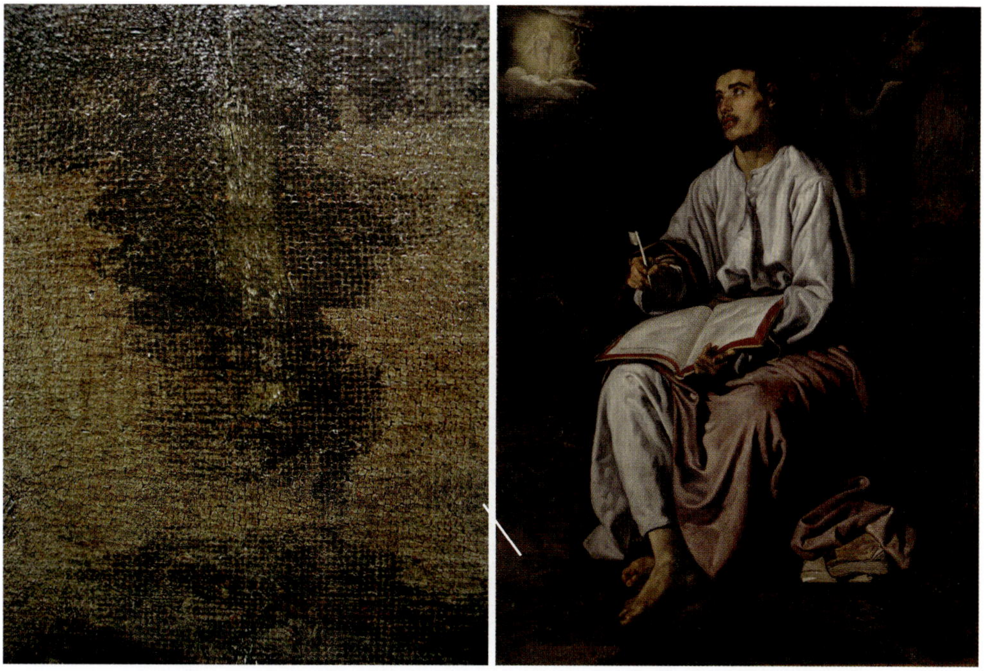

Figure 4 Diego Velázquez, *St John the Evangelist on the Island of Patmos*, *c.*1618, oil on canvas, 135.5 × 102.2 cm, National Gallery, London, NG6264. Detail showing the 'brush-wiping' retouched with ground-coloured paint. (© The National Gallery, London)

the *St John on Patmos*. However, the *St John* also contained some fascinating circumstantial evidence on the original tone and colour of the missing ground. Velázquez had a characteristic and seemingly lifelong habit of testing the colour and loading of his brush by making practice strokes in the margins, which were then covered as he worked; these are often only visible in X-radiographs. In the *St John*, however, he applied some of them in areas where he was otherwise quite satisfied with the colour provided by the priming, and therefore touched-out the rogue brushstrokes with paint mixed to match the priming colour but using other, more stable pigments. While the uncovered priming has blanched, his retouching has remained little changed in its dark olive-brown colour. This is the colour that, when mentally substituted for the blanched areas, starts to make sense of these odd chalky passages (Figure 4). Similar blanching is visible in many paintings from the Seville period, for example in the hand of the servant in the Chicago *Kitchen Scene*.[3] In the case of the *Immaculate Conception* and the *St John* it explained some odd, interesting, but rather minor features. Both paintings remained quite legible in their larger modelling and spatial relationships. As the cleaning progressed on the *Christ in the House of Martha and Mary*, however, it became clear that the exposed ground was significantly more extensive and, therefore, its blanched appearance was more visually disruptive (Figure 5). Velázquez's spatial construction is often somewhat awkward in the early works, resulting in part from the way he assembled his compositions from carefully studied individual elements, as Pacheco describes (Pacheco 1649). Here, the space was made more confused by the disruption of modelling caused by the blanched pigment; we are not sure where the wall is because the shadows cast upon it by fish, carafe and servant are not visible (Figure 6) The recession of

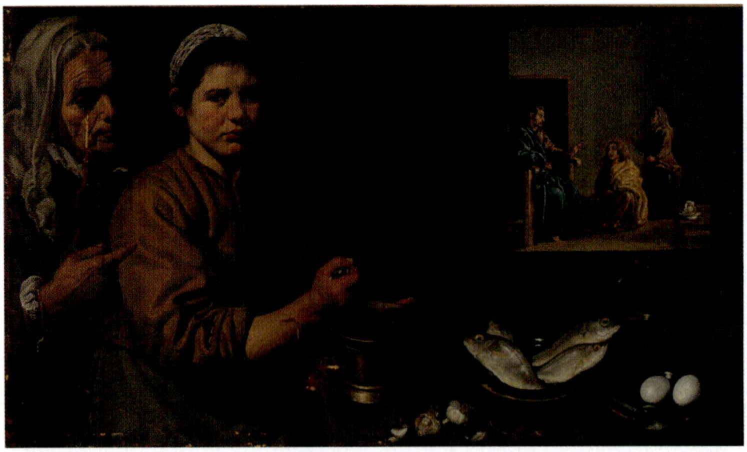

Figure 5 Diego Velázquez, *Christ in the House of Martha and Mary*, NG1375, *c.*1618: after cleaning before retouching. (© The National Gallery, London)

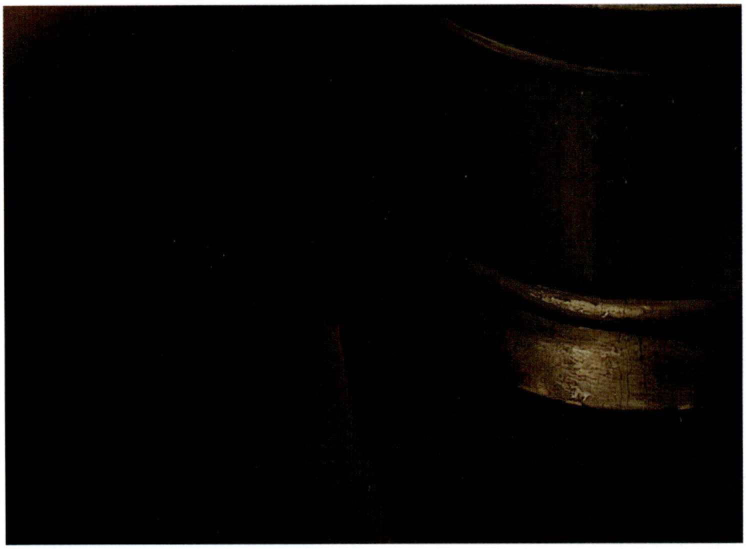

Figure 6 Diego Velázquez, *Christ in the House of Martha and Mary*, NG1375, *c.*1618: detail of the area below the young servant's arm before restoration. (© The National Gallery, London)

the table is confused by the disappearance of the strong cast shadow from the servant's arm, which had become lighter than the table itself, while the rounded modelling of the carafe and the old woman's face had taken on an oddly Cubist flattening as the darker tones provided by the priming colour had become lost.

In this case it was decided to compensate for the pigment change during the retouching campaign, and the chalky areas were glazed down to something approximating their original tonal value and colour (Figure 7). It was not a decision taken lightly, but after much discussion and reflection,[4] nor was it guided by a wish to correct a colour change, but to help restore the intended modelling and spatial relationships. This is the essential point: such toning would not have been considered were it not for

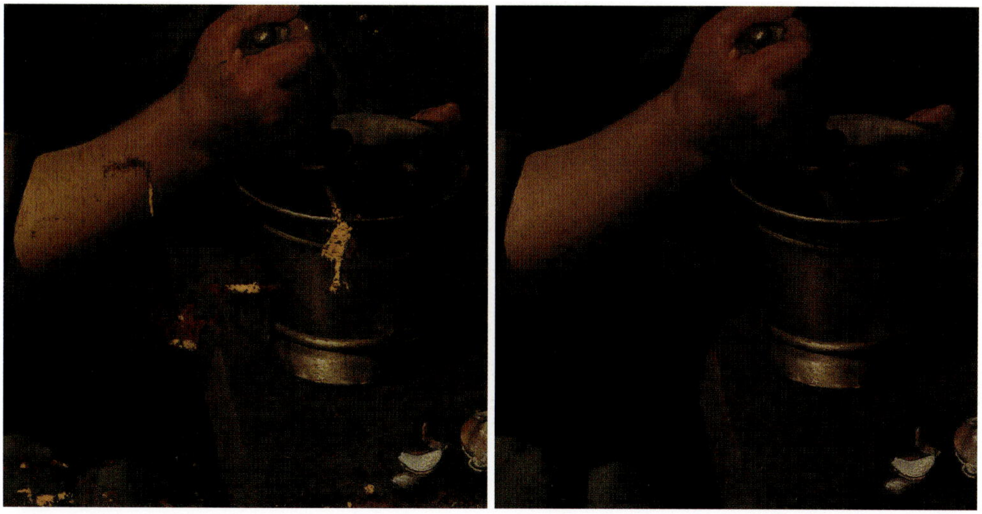

Figure 7 Diego Velázquez, *Christ in the House of Martha and Mary*, NG1375, *c.*1618: detail showing the table before and after retouching. (© The National Gallery, London)

the fact that the altered pigment originally was quite low in chroma or colour intensity. The exposed priming essentially provided a tonal and modelling function, not an intense colour.

Other prominent changes in colour – primarily fading of red lakes – were left completely undisturbed in the restoration. Indeed, simply describing the range of colour change within works by Velázquez could be the subject of another paper. For example, there is ample proof of significant fading of red lakes from all the National Gallery's paintings of the Seville period. Identified as Mexican cochineal,[5] something of their original hue and intensity is suggested in the close observation of tiny areas where the applied glaze has pooled within the brushwork of the underlying body colour, as can readily be seen in the close scrutiny of the *St John* and the *Immaculate Conception*.

Velázquez abandoned the Seville grounds after his permanent move to Madrid, using more stable materials: first a red (essentially iron oxide) preparation and then shifting to a light grey material, rich in lead white. Red lakes remained a constant and important pigment in his palette, however, particularly so for a painter so fascinated by the qualities and effects of 16th-century Venetian painting. Fading of these pigments occurs frequently in works spanning the whole of his career – no doubt based in part on the individual circumstances in which particular works have been displayed – with a range of visual consequences.

It is probably not surprising to learn that the red curtain backdrop and tablecloth within the magnificent *Philip IV in Brown and Silver* have faded significantly, as is clearly indicated by the relatively well-preserved areas around the edge of the painting where they have been protected from light by the frame.[6] Perhaps more surprising, however, is the discovery of the intended colour of the cape and hat, where a combination of discoloured varnish, darkened medium and faded red lakes has shifted the original mauve tonality, which was composed of red lake, azurite, black and white pigments. Close inspection of the costume also shows small flicked strokes of better preserved deep purple paint – probably because no light-scattering white pigment is present within the mixture – which undoubtedly would have been more telling in combination with the original colour of the cape and hat. Also, a comparison of the head of the National Gallery painting with that of the Prado's *Hunting Portrait*[7] – each seemingly based on a tracing

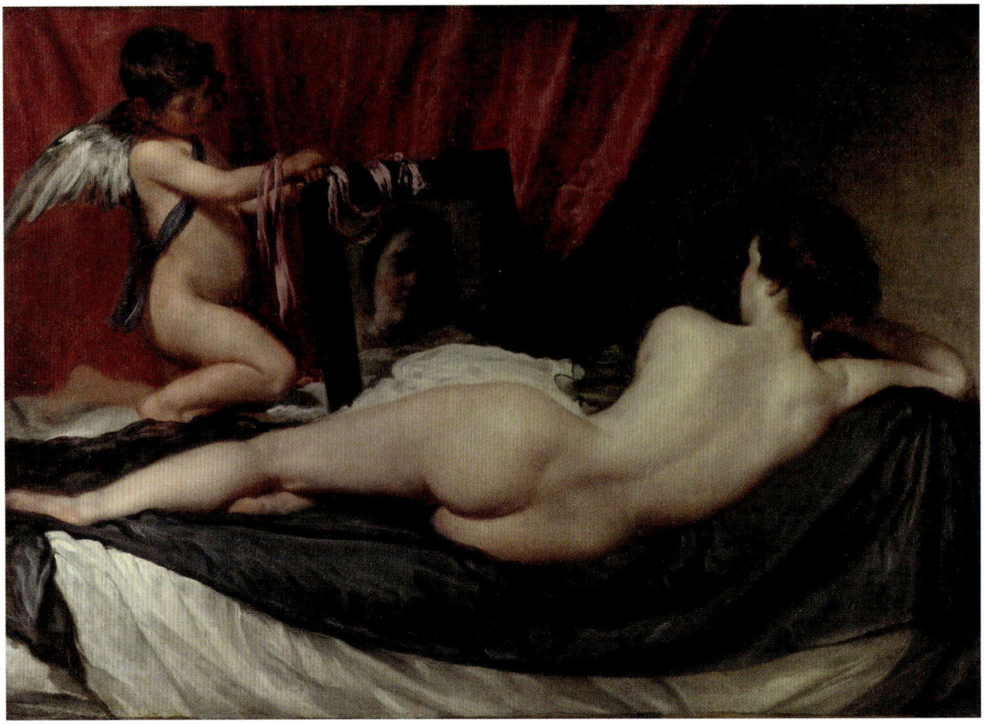

Figure 8 Diego Velázquez, *The Toilet of Venus ('The Rokeby Venus')*, 1647–51, oil on canvas, 122.5 × 177 cm, National Gallery, London, NG2057. (© The National Gallery, London)

of the same lost preparatory study – shows the extent to which the face of the National Gallery painting has also suffered from fading of the red lakes, disrupting the intended modelling of the lips and reducing the general intensity of the flesh tones.

Dating from around the late 1640s, the *Rokeby Venus*[8] (Figure 8) provides another notable example of colour change within paintings by Velázquez at the National Gallery. The painting famously shows an astonishing variety of handling, yet for all that the paint build-ups are simple: for example the red lakes of the backdrop curtain are worked directly on the ground with modelling achieved primarily through simple mixtures applied *alla prima*, rather than superimposed layers of applied colour. These lakes have faded significantly despite the apparent intensity of the curtain, and the loss of colour from this pigment elsewhere has had a serious and surprising effect on our reading of the picture as a whole. The slate grey of the bedcover was originally a deep mauve-purple, the original effect of which must have been richly sensual and quite different from the rather sober chromatic restraint the picture now displays. An image of how the picture might have looked, based on extending tiny areas of relatively well-preserved colour over the larger area in which it was originally applied, was constructed by the National Gallery's Scientific Department (Figure 9).[9] In this context the strong red colour along the contour of Venus's inner thigh is now understandable as a simple reflection of the now-lost colour of the bedcover. It is very interesting to consider this now-lost appearance in comparison with the Prado's nearly contemporary *Coronation of the Virgin* [10] – a painting which remains unlined and appears relatively unfaded, and so sits happily among that imaginary grouping of beautifully preserved condition benchmarks. And

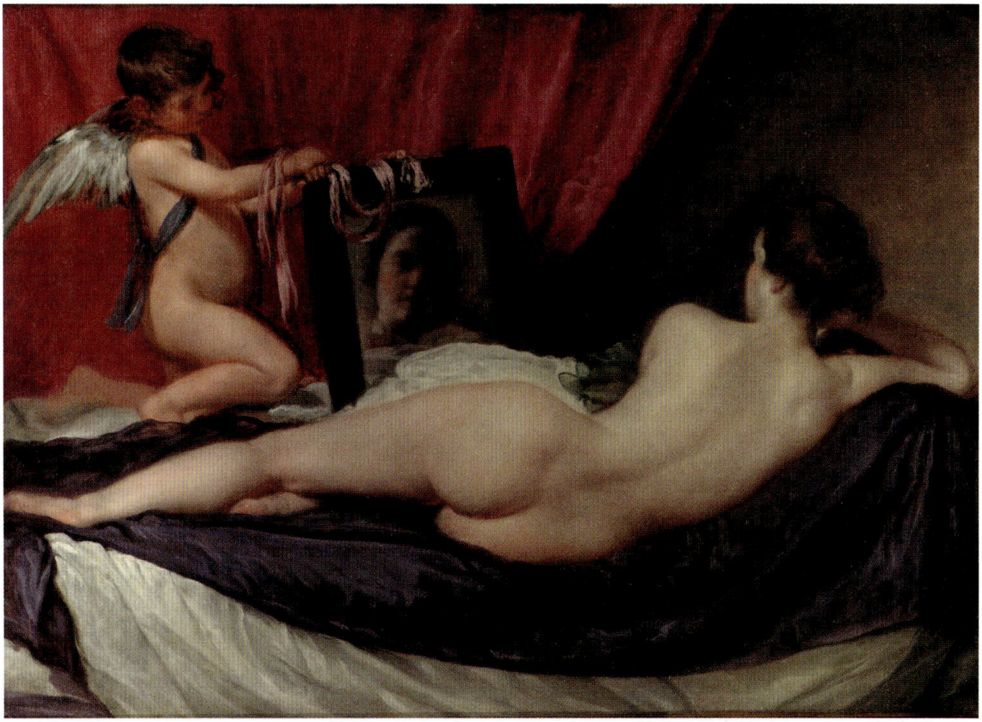

Figure 9 Diego Velázquez, *The Toilet of Venus ('The Rokeby Venus')*, 1647–51: digital reconstruction based on areas of relatively unfaded paint at the edges of the painting. (© The National Gallery, London)

while it is nearly impossible to imagine two works more opposed in subject, each displays a similar formal interest in the rich arrangement of reds, purples and flesh tones. In a more general sense, a revised appreciation of the intended colouring of the National Gallery painting adds something significant to the way Velázquez might be considered as a colourist, for example, as part of the larger discussion of his long and deep engagement with Titian and Tintoretto.

However, as interesting and informative as the digital reconstruction of the *Rokeby Venus* may be, there was never any consideration given to trying to reverse these colour changes within an actual restoration. No one could reasonably argue that the fading has caused a fundamental distortion or confusion of the larger spatial construction as was the case with *Martha and Mary*, however significant the loss of colour.

In the treatment of the *Martha and Mary*, no attempt was made to redress the lost chromatic intensity of the faded red lakes. There was no intention of making the painting's colour look as it did originally – the glazing of the blanched priming was simply intended to ease the reading of Velázquez's composition. The red lakes remain faded, the generally increased transparency remains unaltered, and the painting is forever affected by the historic abrasion over much of its surface. Some aspects have become more clear and understandable: the table is now understood as being placed against the wall and the biblical scene is more likely seen through an aperture in the wall than an image displayed upon it. However, the improved understanding of the closeness of the wall to the servant girl also emphasises the impossibility of the old woman actually fitting her left shoulder where it is depicted. It is the sort of awkwardness that seems to speak to that additive compositional method described by Pacheco (Keith 2006; Veliz 1996), perhaps together with the formal influence of prints on the young artist. It is, in any case, forgivable from a painter still in his teens, and it remains an awkward feature that cannot be explained by any sort of change in condition and is therefore unaffected by the recent treatment. However that awkwardness might be explained, it was our intention that all may now consider it based on a more secure reading of the picture.

Such an intervention is essentially about helping with the 'reading' of paintings, which is difficult to describe but of fundamental importance. Museum restorers by training and experience have ample opportunity to consider art-historical ways of understanding the works with which they are charged, elements including iconography, stylistic features as well as materials and techniques, but sometimes it can also be helpful to step back a little and look in other ways. It is fascinating to consider examples of how later artists described certain Old Master paintings, and the features they have found impressive, such as Van Gogh's appreciation of the simultaneous contrasts of blue and orange within Frans Hals' Rijksmuseum Militia Company portrait, the so-called *Meagre Company*.[11] Turner's notes on Guido Reni's *Aurora* are wholly, and sensitively, concerned with its colour relationships, giving little or no consideration to subject or formal composition.[12]

Such responses are fascinating examples of interest in the formal qualities of paintings that in a sense lie outside of art history – whether composition, tone or colour – perhaps even condition. They also serve to remind us of the wide range of highly engaged responses that such works can evoke. There is no doubt that cleaning and restoration of paintings often results in significant changes in appearance, but we remain very reluctant to intervene where changes in colour are concerned, restricting ourselves to the very exceptional correction of largely tonal values when the basic compositional organisation has been significantly compromised. Scientific research continues to provide more information on the nature and degree of changes which can occur, allowing better understanding of original intent and appearance. But these changes, however significant, still leave us with paintings of great impact and beauty and so do little to blunt the more timeless qualities of great works of art.

Notes

1. Carlo Dolci, *Adoration of the Kings*, 1649, National Gallery (NG6523), 117 × 92 cm; Guido Reni, *The Coronation of the Virgin*, *c*.1607, National Gallery (NG214), 66.6 × 48.8 cm; Peter Paul Rubens, *Samson and Delilah*, *c*.1609–10, National Gallery (NG6461), 185 × 205 cm.
2. Scientific investigation was led by Ashok Roy, who also was responsible for cross-section analyses and pigment identification, while medium analyses were done by Rachel Morrison.
3. Diego Velázquez, *Kitchen Scene*, 1618–20, Art Institute of Chicago (Robert Walker Memorial Fund, 1935.380), 55.9 × 104.2 cm. I would like to thank Frank Zuccari, Executive Director of Conservation, Art Institute of Chicago, for allowing close viewing and sharing his invaluable knowledge of the painting.
4. These discussions took place in close collaboration with the painting's curator, Dr Dawson Carr, whose wide knowledge of conservation issues and expertise on Velázquez was essential to the process.
5. Identified by the National Gallery Scientific Department using high pressure liquid chromatography (HPLC).
6. *Philip IV of Spain in Brown and Silver*, NG1129, *c*.1631–2; Keith 2006.
7. Diego Velázquez, *Philip IV in Hunting Dress*, 1632–34, Museo Nacional del Prado (inv. no. P01184), 189 × 124 cm.
8. Velázquez, *The Toilet of Venus*, NG2057, 1647–51; Keith 2006.

9. The digital reconstruction was made by Joseph Padfield.
10. Diego Velázquez, *The Coronation of the Virgin*, 1635–36, Museo Nacional del Prado (inv. no. 1168), 178.5 × 134.5 cm.
11. Frans Hals, Pieter Codde, *Militia Company of District XI under the Command of Captain Reynier Reael, known as 'The Meagre Company'*, 1637, Rijksmuseum (inv. SK-C-374), 233 × 452 cm. Van Gogh describes it in a letter to his brother Theo, 10 October 1885: 'In itself, that painting alone makes the trip to Amsterdam well worth while, especially for a colourist. There's a figure in it, the figure of the standard-bearer in the extreme left corner, right up against the frame. That figure is in grey from top to toe, let's call it pearl grey, – of a singular neutral tone – probably obtained with orange and blue mixed so that they neutralize each other – by varying this basic colour in itself – by making it a little lighter here, a little darker there, the whole figure is as it were painted with one and the same grey. But the leather shoes are a different material from the leggings, which are different from the folds of the breeches, which are different from the doublet – expressing different materials, very different in colour one from another, still all one family of grey – but wait! Into that grey he now introduces blue and orange – and some white. The doublet has satin ribbons of a divine soft blue. Sash and flag orange – a white collar. ... Orange, white, blue, as the national colours were then. Orange and blue next to each other, that most glorious spectrum – on a ground of grey judiciously mixed, precisely by uniting just those two, let me call them poles of electricity (in terms of colour, though) so that they obliterate each other, a white against that grey. Further carried through in that painting – other orange spectrums against a different blue, further the most glorious blacks against the most glorious whites – the heads – some twenty– sparkling with spirit and life, and how they're done! and what colour! the superb appearance of all those fellows, full length. But that orange, white, blue chap in the left corner ... I've seldom seen a more divinely beautiful figure – it's something marvellous. ... Delacroix would have adored it — just adored it to the utmost.'
12. Turner on Guido Reni, from a notebook of 1819, folio 4r (D16766), Tate, Turner Bequest CXCIII3 (www.tate.org.uk/art/research-publications/jmw-turner/joseph-mallord-william-turner-notes-on-the-aurora-by-guido-reni-in-the-palazzo-pallavicini-r1138541): The Aurora of Guido – the sky rather | yellow like W Smoke Light on a Dark | red ground on which [?G] rather stiks his | Relievo excepting the orange drapery the | light Lilac serves to detach the figure | but at the same time takes away all | aerial quality from the sky which | perhaps is too solid the Dark figure | with golden Hair and light Blue Drapery | is beautiful as to form, color and | design the Drapery is the lightest colour | in the picture. the greenish robe | of the next figure second the next a lilac | robe and light drapery the Apollo red Lake | and comes off whole .: half tone from the | sky the half figures one green the other | red serves to clear the [sketch of figure] [?figures] the | Sea is dark Blue and rather aerial by [?reason]2)

References

Carr, D. *et al.* (2006) *Velázquez*, exh. cat. London: National Gallery.

Christiansen, K., Kanter, L. and Strehlke, C. (1988) *Painting in Renaissance Siena, 1420–1500*, exh. cat. New York: Metropolitan Museum of Art.

Garrido Perez, C. (1992) *Velázquez: tecnica y evolucion*. Madrid: Museo Nacional del Prado.

Keith, L. (2006) 'Velázquez's painting technique', in D. Carr *et al.*, *Velázquez*, exh. cat. London: National Gallery, 70–92.

Pacheco, F. (1649) *Il arte de la pintura*. Seville.

Syson, L. and Gordon, D. (2001) *Pisanello: Painter to the Renaissance Court*, exh. cat. London: National Gallery.

Taylor, P. (2015) *Condition: The Ageing of Art*. London: Paul Holberton Publishing.

van Eikema Hommes, M. (2004) *Changing Pictures: Discoloration in 15th–17th Century Oil Paintings*. London: Archetype Publications.

Veliz, Z. (1996) 'Velázquez's early technique', in D. Davies (ed.), *Velázquez in Seville*, exh cat. Edinburgh: National Galleries of Scotland.

Author's address

Larry Keith, Head of Conservation, National Gallery, London, UK (larry.keith@btinternet.com)

Colour changes due to the fading of Prussian blue in Danish Golden Age paintings

Troels Filtenborg, David Buti, Anna Vila and Jørgen Wadum

Introduction

The early history, manufacture and properties of Prussian blue (a hexacyanoferrate(II) blue pigment) have been widely studied and discussed. One of the most significant additions to artists' colours in the 18th century, the pigment is a marker for the emerging role of chemists in the evolution of painting materials. It has been demonstrated by experimental studies and is known from historical sources that, due to its high tinting strength, Prussian blue was often used – mostly in moderate proportions – mixed with other compounds and pigments. Among its virtues, the relatively modest price would seem to account for its frequent occurrence in paintings from the mid-18th century onwards. Following its discovery, production of Prussian blue was taken up rapidly at a number of different locations in Europe, although reservations concerning its permanence also quickly appeared in technical writings. As early as the mid-18th century, several sources claimed that the pigment was not entirely stable, and in his *Handmaid to the Arts* from 1758, Robert Dossie famously states: 'The common Prussian blue, found in the shops, which is prepared almost wholly at present by ignorant and sordid people, and sold at very low prices, can be very little depended upon in paintings of consequence' (Dossie 1758: 77–8). The poor reputation of the pigment persisted well into the following century: in 1827 Pierre Louis Bouvier, in his *Manuel des jeunes artistes et amateurs en peinture*, referred several times to the impermanence of Prussian blue claiming, however, that the pigment manufactured in England is the more stable variety (Bouvier 1827: 32 n. 1, 39, 43–5). Jean-François-Léonor Merimée, in his almost contemporary *De la peinture à l'huile* from 1830, also describes the English method of preparation, stating that it has for a while been the only one in use (Merimée 1830: 177–8).[1]

Studies by Kirby and Saunders and by Samain *et al.* have indicated the link between the fading and colour changes of Prussian blue and the different 18th- and 19th-century methods of its preparation (Kirby 1993; Kirby and Saunders 2004; Samain *et al.* 2013a). Furthermore, the experimental studies conducted by Kirby and Saunders as well as Samain showed that the permanence of the blue was affected by the addition of white pigment which increased the irreversible colour change recorded on exposure to light (Kirby and Saunders 2004; Samain *et al.* 2013b). They concluded that the principal factor in the stability of different varieties of Prussian blue, historical as well as modern, was indeed the presence of extenders in the blue pigment, either incidental due to inadequate purification or as a result of deliberate addition of, for instance, a white pigment. The strong tendency of modern grades of Prussian blue to fading when mixed with white in a linseed oil medium has further been demonstrated by Samain *et al.* (2011), who established that the nature of the white pigment is a determining factor in the process. The degree of fading was shown to increase strongly when Prussian blue was mixed with lead white or zinc white, the hypothesis being that the manifest light sensitivity may be related to the accelerated oxidative polymerisation of the linseed oil in the presence of pigments that tend to form soaps with the free acids in oil.

In Denmark, and by and large for easel painting this would mean Copenhagen, the earliest documented use of Prussian blue is by 1721 (Ludvigsen *et al.* 2014). The pigment remained in widespread use throughout the 18th as well as the 19th century, so reservations about its permanence evidently did not restrain Danish painters from employing it. Part of the explanation may be that while there was among writers some knowledge of the problems with the permanence of Prussian blue and the fact that it could be manufactured by various methods, resulting in differing compositions, it is debatable whether this awareness was shared by users of the pigment. No positive records of any domestic production have emerged, so in all probability the pigment sold in Denmark was imported, and given that suppliers in Copenhagen could have obtained their product from a number of foreign sources, the qualities available to their clients are likely to have varied. The same probably applied to the different grades sold under the name Prussian blue or any of the synonyms under which it was marketed. There are no records of specific 'colourmen' or artists' supply shops operating in Copenhagen during this period; they seem to emerge only by the second half of the century (Bregnhøi 2010: 157–60). Until then, grocers were traditionally the main suppliers of pigments. However, the market for artists' materials in a town of a modest size such as Copenhagen in the early decades of the 19th century would have been limited and able to support only a quite narrow circle of traders. This meant that artists sometimes depended on other sources for their materials. Christoffer W. Eckersberg, for example, according to his diary and a memoir by his daughter, regularly purchased pigments from the house painter L.E. Ernst (Villadsen 2009, 2: 912, 939, 979, 980, 1002, 1052, 1214; Eckersberg 1917: 21), who is likely to have stocked materials in grades intended predominantly for his own enterprise and perhaps not always of a quality entirely suitable for artists' use.

It is altogether doubtful whether dealers in Copenhagen were concerned about the exact nature of their product or if artists were, in general, qualified to assess the importance of the pigment grade or the fact that the durability of the colour could change dramatically when the pigment was mixed with certain whites. Not surprisingly, a degree of fading of the blue in such mixtures has indeed been found at regular intervals in Danish paintings from the 18th century and even more often from the 19th. The following cases, exemplifying the issue, all date from the first half of the 19th century which was a highly fertile period in Danish art. Technical examination of the paintings – by Nicolai A. Abildgaard, Christoffer W. Eckersberg and Christen Købke – revealed the presence of Prussian blue mixed with lead white as a component in the affected areas.

Nicolai A. Abildgaard, *Pamphilus and his Servant Davus: A Scene from Terence's Comedy 'The Girl from Andros'* (1802)

The painting (Figure 1b) was executed by the painter, architect and designer Nicolai Abildgaard (1743–1809) for his own home as the second in a series of four paintings depicting scenes from Terence's comedy *The Girl from Andros* (Figure 1).[2] Visual examination revealed a marked fading of the blue paint in the sky, evident when comparing the preserved colour along the edges with areas that had not

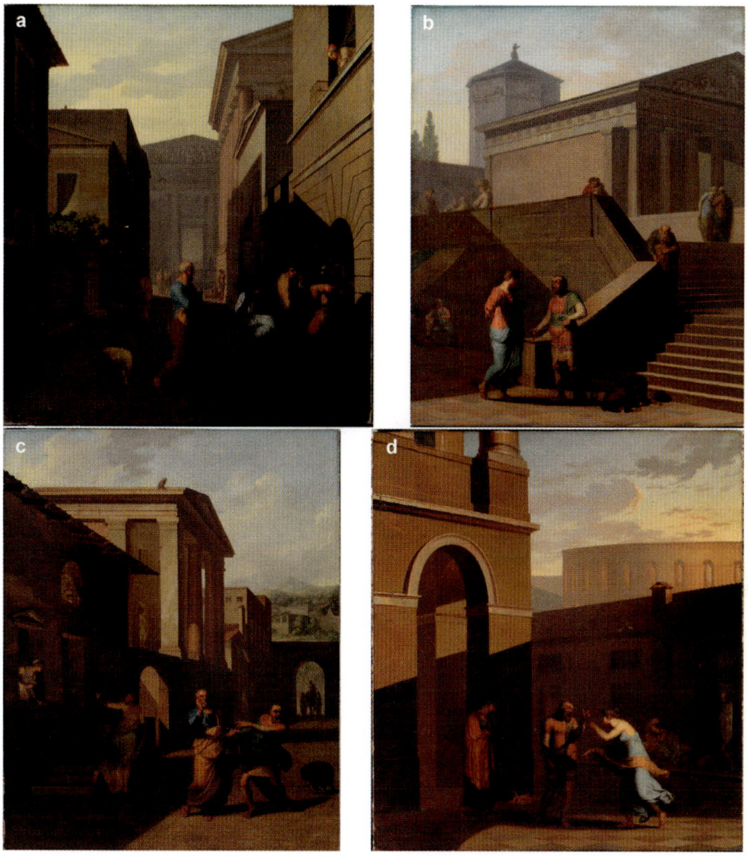

Figure 1 (a) Nicolai Abildgaard, *Simo and his Former Slave Sosia*, 1803, oil on canvas, 157.5 × 142 cm, Statens Museum for Kunst, Copenhagen, KMS587. (b) Nicolai Abildgaard, *Pamphilus and his Servant Davus*, 1802, oil on canvas, 157.5 × 128.5 cm, Statens Museum for Kunst, Copenhagen, KMS588. (c) Nicolai Abildgaard, *The Midwife Taking Leave of the Girl from Andros*, 1801, oil on canvas, 157.5 × 128.5 cm, Statens Museum for Kunst, Copenhagen, KMS589. (d) Nicolai Abildgaard, *The Servant Davus and the Maid Mysis*, 1804, oil on canvas, 157.5 × 142 cm, Statens Museum for Kunst, Copenhagen, KMS590. (Photos © SMK Foto)

been protected from the light by the rebate of the frame (Figure 2). Similar fading was seen in the sky of the painting depicting the third scene of the comedy (Figure 1c), although in that case it was a little less pronounced.

An X-radiograph of *Pamphilus and his Servant Davus* shows that Abildgaard – as he did in the other three scenes – was careful in considering the position of the sun and the angle of the light when, among other changes during the execution, he extended the shadow cast by the architecture behind the two principal characters in the left-hand foreground. As in most plays by Terence, the plot of *The Girl from Andros* is played out over one single day from morning to evening. The widely read Abildgaard, with his extensive knowledge of classical literature, would have been aware of this particular feature, choosing to emphasise it in the way he depicted the light and the sky in the four scenes. This would therefore have been a much more differentiated and prominent visual element of the series in its original state, but is today far less evident as a result of the fading of the skies in the two middle scenes.

Figure 2 Nicolai Abildgaard, *Pamphilus and his Servant Davus*: detail from the left-hand edge of the sky showing fading of the blue colour.

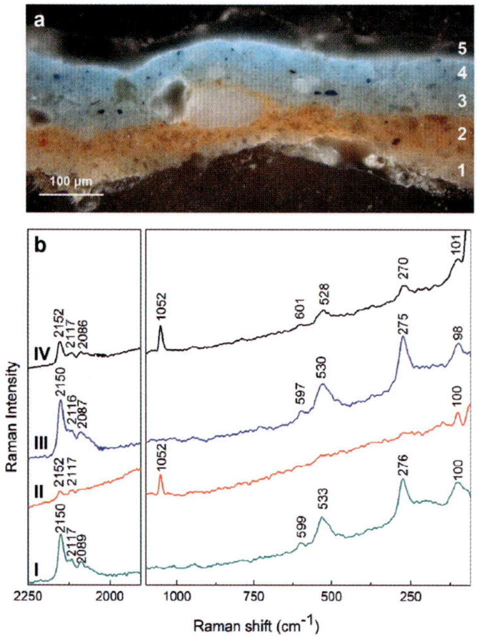

Figure 3 Nicolai Abildgaard, *Pamphilus and his Servant Davus*. (a) Cross-section of the ground and paint layers at the left-hand edge of the sky where the paint has been protected from light by the rebate of the frame. (b) Raman spectra obtained in the blue layers: I (dark green line) and II (red line) recorded on paint layer 3; III (blue line) and IV (black line) recorded on paint layer 4. The spectra show the characteristic bands of Prussian blue and lead white. Spectra have been vertically translated.

It appears that Prussian blue was the only blue pigment Abildgaard ever used: no other blues have been identified by examination of a large number of his paintings from throughout his career (Filtenborg 2014: 54–7). Prussian blue was of course an attractive alternative to ultramarine ($Na_{8-10}Al_6Si_6O_{24}S_{2-4}$), for instance, due to its high tinting strength and modest price. However, if he consulted Pernety's *Dictionnaire portative de peinture, sculpture et gravure* and Constant de Massoul's *A Treatise on the Art of Painting* from his own extensive library, Abildgaard would have been conscious of the problems with the permanence of the pigment (Pernety 1797: xciii; de Massoul 1797: 182–3). What is certain is that the potential awareness of the issue did not make him abandon Prussian blue.

A cross-section of a sample taken from the sky under the frame[3] is shown in Figure 3a. The paint layer of the sky appears to be composed of two blue layers, with the colour of the top layer slightly brighter than the one beneath it. Raman spectra recorded from the two blue layers[4] reveal the presence of Prussian blue (Figure 3b). Both spectra are characterised by signals at 275, 530 (probably attributable to the stretching and bending modes of Fe–C, Fe–C–N, C–Fe–C), 2087, 2117 and 2150 cm^{-1} (CN^- stretching) (Samain *et al.* 2013a). Furthermore, both layers are also characterised by a Raman band at 1052 cm^{-1} (attributed to the symmetric stretching of $CO_3^{=}$) (Brooker *et al.* 1983) indicating the presence of lead white in the blue paint mixture.

Christoffer W. Eckersberg, *The Corvette 'Galathea' in a Storm in the North Sea* (1839)

In contrast to Nicolai Abildgaard, who seems to have employed Prussian blue as his sole blue colorant, examination of the paintings by Christoffer W. Eckersberg (1783–1853) shows a different pattern in the occurrence

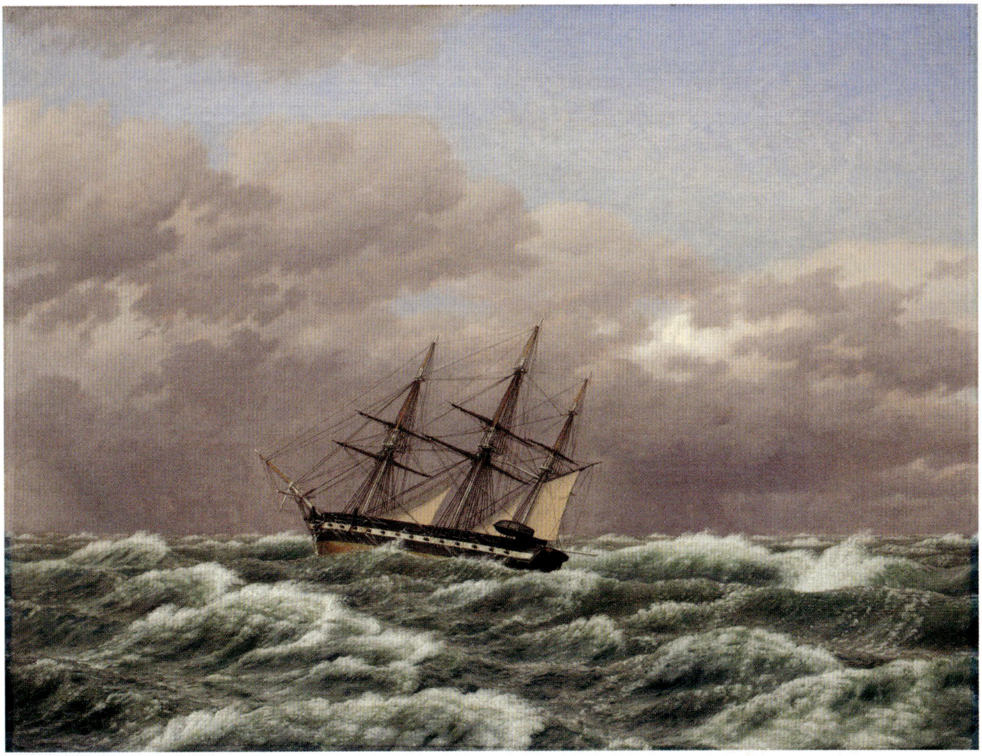

Figure 4 Christoffer Eckersberg, *The Corvette 'Galathea' in a Storm in the North Sea*, 1839, oil on canvas, 47.5 × 63.5 cm, Statens Museum for Kunst, Copenhagen, KMS1116. (© SMK Foto)

of the pigment. Eckersberg worked in Paris and Rome from 1810 to 1816, and he appears to have preferred ultramarine (an alumino-silicate containing sodium ions and ionic sulphur groups) and cobalt blue (a cobalt aluminium oxide) during this sojourn, perhaps because of the greater availability of those pigments in the two cities. In fact he was one of the early users of cobalt blue, as seen in his accounts which show that he purchased some in 1812 (commercial production of the pigment began in France around 1807).[5] Paintings by him in which Prussian blue have been identified are dated to the period before he left Copenhagen and after he returned in 1816. It is these paintings which have obvious colour changes related to the pigment. One of them, Eckersberg's depiction of *The Corvette 'Galathea' in a Storm in the North Sea* (Figure 4), painted in 1839, was based on extensive notes made in his diary, as he was in fact on board the vessel itself on a voyage to Dover. The diary entry vividly describes the sea and the storm which occurred in May (whereas he started on the painting a month later), and he writes that when the sun came out, the sea had beautiful pure blue and green tones.[6]

As the artist who is credited with introducing *plein air* studies in Danish art, Eckersberg's painting is characterised by an insistent truthfulness, not least in his recording of weather conditions. He pursued this quality in his representations, which is perhaps particularly evident in his many marine paintings, including the depiction of the *Galathea*. However, examination of the painting shows that the paint of the sea has preserved the colour originally intended by Eckersberg only along the edges where it has not been exposed to the

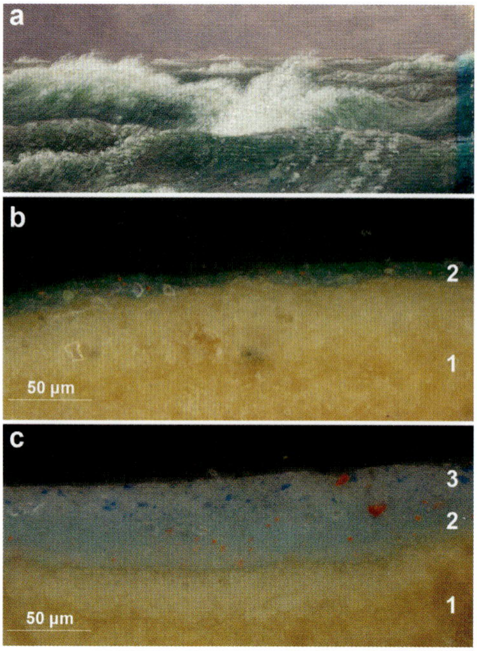

Figure 5 Christoffer Eckersberg, *The Corvette 'Galathea' in a Storm in the North Sea*. (a) Detail from the right-hand edge of the sea showing fading of the greenish-blue colour. (b) Cross-section of ground and paint layers from the right-hand edge of the sea where the paint has been protected from light by the rebate of the frame. (c) Cross-section of ground and paint layers from the right-hand edge of the sky where the paint has been covered by the rebate of the frame.

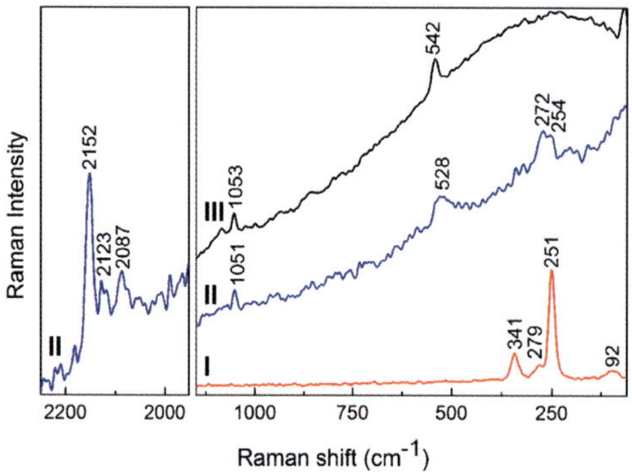

Figure 6 Christoffer Eckersberg, *The Corvette 'Galathea' in a Storm in the North Sea*: Raman spectra obtained in the blue paint layers of the cross-section from the non-exposed area of the sky (Figure 5c). I (red line) spectrum recorded on a red particle in layer 2 showing the characteristic bands of vermilion; II (blue line) spectrum recorded on a blue particle in layer 2 showing the characteristic bands of Prussian blue, lead white and vermilion; III (black line) spectrum recorded on a blue particle in layer 3 showing the characteristic bands of ultramarine and lead white. Spectra have been vertically translated.

light, whereas the overall appearance of the water has changed dramatically from what he experienced at this point of his voyage, owing to the fading of the blue component (Figure 5a). By comparison, the colour of the sky in the painting has not been affected to any significant degree, and cross-sections from the sea and the sky,[7] respectively, give an indication of the reason (Figures 5b and 5c). The sample from the sea has a thin blue-greenish paint layer (*c*.10 μm thick) containing a very fine-grained blue pigment with the appearance of Prussian blue, mixed with a yellow compound and some white and red particles. The sample from the sky has a layer of small heterogeneous blue particles, with the morphology and colour of ultramarine, mixed with a white pigment and a few red pigment particles on top of two blue layers containing a very fine-grained blue pigment, characteristic of Prussian blue, mixed with white and scattered small red pigment particles.

Figures 6 and 7a show the Raman spectra recorded on cross-sections from the sky and the sea.[8] The heterogeneous blue particles in the top blue layer of the sky were confirmed to be ultramarine (Raman band at 542 cm^{-1} ascribed to the stretching of S_3^-) (Figure 6, blue line) (Clark and Franks 1975). In both cross-sections, the signals at *c*.275, 528, 2087, 2120 and 2150 cm^{-1} are those of Prussian blue, while that at 1050 cm^{-1} is attributed to lead white. The assignments of their Raman bands are reported in the previous section. Furthermore, the Raman scattering bands of vermilion, assigned to the bending and stretching of Hg-S (Frost *et al.* 2002), are also visible at 92, 250, 275 and 340 cm^{-1} (Figures 6 and 7a, red lines). No Raman bands ascribable to a yellow component are present in the collected spectra on the sample from the sea. However, X-ray fluorescence (XRF) measurements[9] carried out on the same area show, in addition to the predominant signals of lead, those of antimony and tin (Figure 7b) together with iron, barium, mercury, strontium and calcium. It is possible that the artist used either a mixture of Naples ($Pb_2Sb_2O_7$) and lead-tin (Pb_2SnO_4 and/

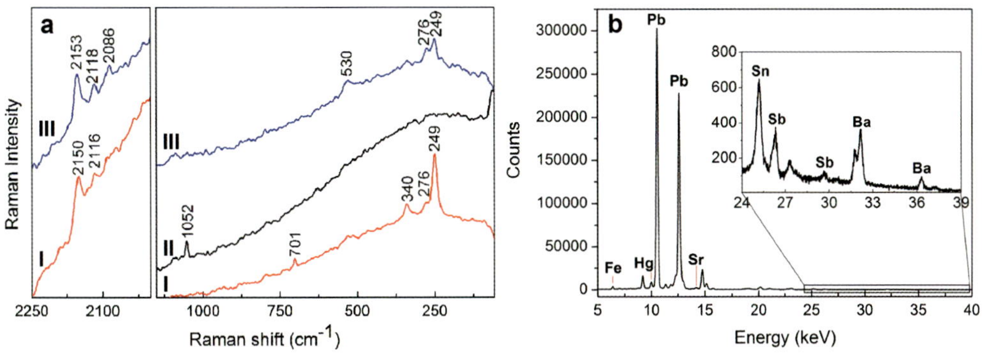

Figure 7 Christoffer Eckersberg, *The Corvette 'Galathea' in a Storm in the North Sea.* (a) Raman spectra obtained in the greenish-blue paint layer of the cross-section from the non-exposed area of the sea (Figure 5b). I (red line) spectrum recorded on a red particle in layer 1 showing the characteristic bands of vermilion and Prussian blue; II (black line) spectrum recorded on a white particle in layer 1 showing the characteristic band of lead white; III (blue line) spectrum recorded on a blue particle in layer 1 showing the characteristic bands of Prussian blue and vermilion. Spectra have been vertically translated. (b) XRF spectrum recorded in a non-exposed area of the sea. The strongest peaks are characteristic of lead (Pb) while mercury (Hg) and iron (Fe) are present in low amounts. The inset shows the presence of very low amounts of antimony (Sb), tin (Sn) and barium (Ba).

or $Pb_2Sn_{1-x}Si_xO_3$) yellows, or a Pb-Sn-Sb triple oxide yellow (Hradil *et al.* 2007; Sandalinas *et al.* 2006).

Christen Købke, *View of Lake Sortedam from Dosseringen Looking Towards the Suburb Nørrebro outside Copenhagen* (1838)

Købke's painting (Figure 8), almost contemporary with Eckersberg's *Galathea*, has traditionally been seen as the culmination of a group of his works from the second half of the 1830s in which he introduced a dreamier and more poetic atmosphere (Vila *et al.* 2014; Wadum *et al.* 2010). It shows Lake Sortedam in what was then the outskirts of Copenhagen, and the so-called embankment, on what appears to be a quiet summer evening. The water of the lake has a distinct pale reddish tone, presumably from the sunset, whereas the sky is dominated by a pale bluish colour. There has been much speculation about the painting, partly perhaps due to its lack of obvious subject matter or narrative, and it has been suggested that its specific mood was expressed by a particularly subtle use of colour in which Købke 'remoulded the world of natural phenomena' (Levey 1989; see also Schwartz 1992; Wivel 1993: 36–7; Monrad and Conisbee 1993: 161–6; Monrad 1996; Gunnarsson 1998: 30–31; Berman 2007: 119 ff.; Nørregård-Nielsen 1996, 2: 187–94; Jackson and Monrad 2010: 82–5).

The painting was a carefully considered work, preceded by a large number of preliminary drawings as well as an oil sketch on paper. One of the drawings has inscriptions with notes on colours, sunlight, atmospheric effects and reflections. Interestingly, there are no clues among the notations as to the original appearance of the sky, and no inscription indicating the colour of the lake, as one would expect to find if those had been in some way extraordinary. The small oil sketch on paper, presumably executed on the spot as a record to help Købke's memory when he later undertook the painting in his studio, shows no particular differences between the colour of the sky and the lake. In the sketch Købke chose the lighting as it is around noon, with the sunlight coming from the left, a feature he repeated in the final painting. Here, however, the traditional assumption is that he replaced the sunny day of the study with a more atmospheric and

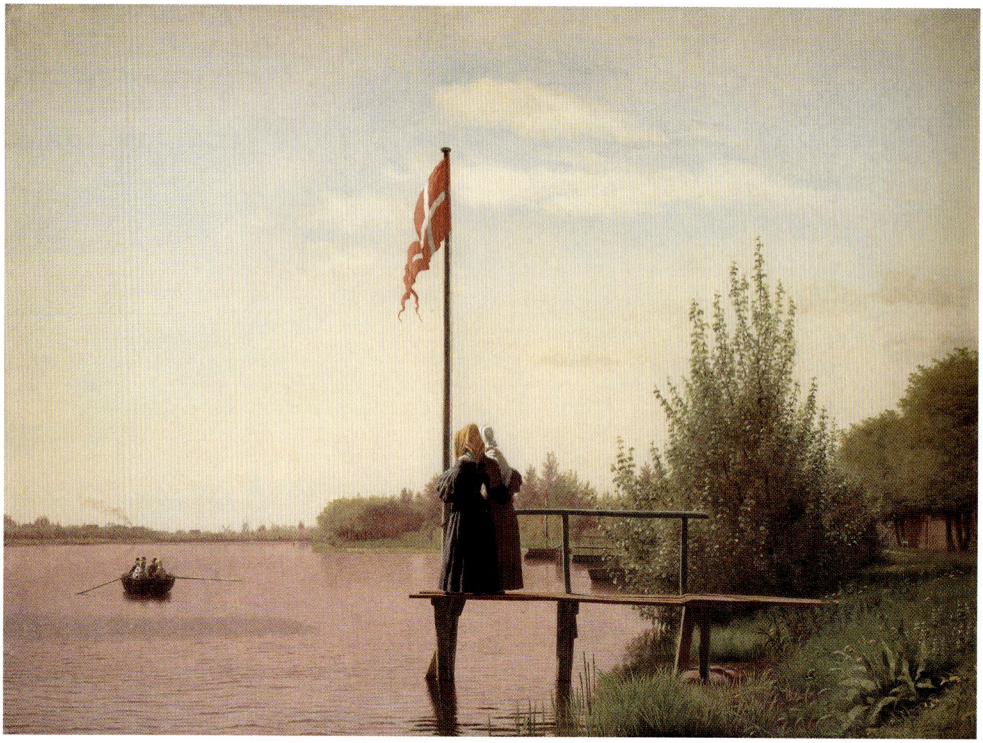

Figure 8 Christen Købke, *View of Lake Sortedam from Dosseringen Looking Towards the Suburb Nørrebro outside Copenhagen*, 1838, oil on canvas, 53 × 71.5 cm, Statens Museum for Kunst, Copenhagen, KMS359. (© SMK Foto)

poetical evening, shortly before sunset, and thereby introduced clearly romantic tones into the picture (but due to an assumed error or oversight, kept the light coming from the left).

However, examination of the edges of the painting revealed that fading has occurred in the blue component of the sky as well as the lake (Figures 9a and 10a). Samples from the areas of the lake and sky[10] which had, respectively, been exposed to and protected from light, show a simple structure with no more than one or two paint layers on top of the ground. Furthermore, they show a paint layer with lead white as the dominant pigment. In the samples from the lake (Figures 9b and 9c), the pale blue paint has scattered red particles of vermilion (Raman bands[11] at 100, 250, 280 and 340 cm^{-1} and confirmed by the presence of Hg by XRF[12]) which, upon the fading of the blue, has become dominant in the tonality of the water despite the presence of Prussian blue with Raman bands at 2090, 2120 and 2152 cm^{-1} and lead white around 1048 cm^{-1} (Raman spectra shown in Figure 9d).[13] In comparison, the red pigment appears to be absent in the samples from the sky where the blue of the exposed area of the sky (Figure 10b) has almost entirely lost its tone as compared with its equivalent from the paint under the rebate of the frame (Figure 10c), which has a more distinctly blue tone. Raman spectra from both the protected and exposed areas indicate the presence of Prussian blue mixed with lead white (Figure 10d).[14]

In view of the pigment changes that have occurred in the paint layers, the original appearance of Købke's depiction of Lake Sortedam appears to have been a more 'natural' one, i.e. with a less striking discrepancy between the colours of the sky, which would initially have been more intensely blue, and the water that would have been more greyish blue

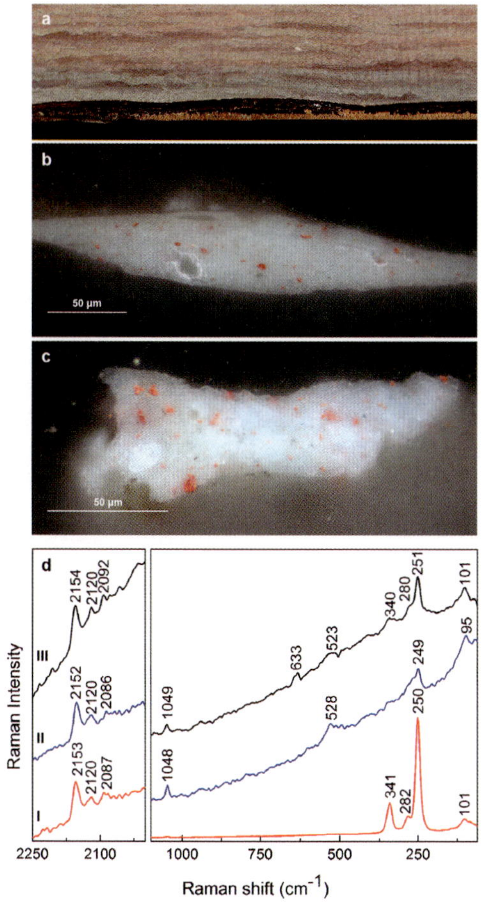

Figure 9 Christen Købke, *View of Lake Sortedam from Dosseringen Looking Towards the Suburb Nørrebro outside Copenhagen*. (a) Detail of the bottom edge of the water where the paint has been protected from light by the frame. (b) Cross-section of the paint layer sampled in the lake in an area exposed to the light. (c) Cross-section of the paint layer sampled at the bottom edge of the lake where the paint has been protected from light. (d) I (red line) Raman spectrum recorded on a red particle in the sample from the exposed area showing the characteristic bands of vermilion and Prussian blue; II (blue line) Raman spectrum recorded on a blue particle in the sample from the exposed area showing the characteristic bands of Prussian blue, vermilion and lead white; III (black line) Raman spectrum recorded on a blue particle in the sample from the non-exposed area showing the characteristic bands of Prussian blue, vermilion and lead white. Spectra have been vertically translated.

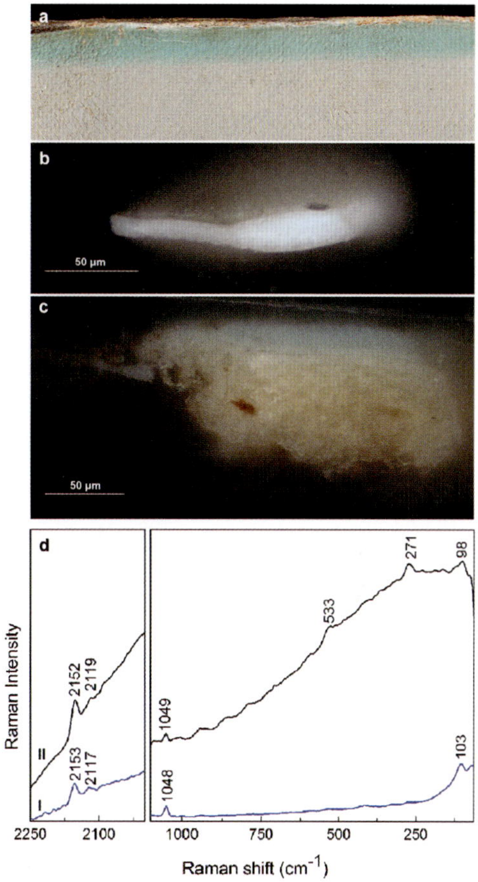

Figure 10 Christen Købke, *View of Lake Sortedam from Dosseringen Looking Towards the Suburb Nørrebro outside Copenhagen*. (a) Detail of the right-hand edge of the sky where the paint has been protected from the light by the frame. (b) Cross-section of the paint layer in the sky in an area exposed to the light. (c) Cross-section of ground and paint layers at the top edge of the sky where the paint has been protected from the light. (d) I (blue line) Raman spectrum recorded on a blue particle in the sample from the exposed area showing the characteristic bands of Prussian blue and lead white; II (black line) Raman spectrum recorded on a blue particle in the sample from the non-exposed area showing the characteristic bands of Prussian blue and lead white. Spectra have been vertically translated.

rather than purplish. This suggests a different reading of the image, which instead of representing the previously supposed evening or late afternoon scene, was originally a painting of the scenery in broad daylight around noon on a quiet summer's day.

Conclusions

Blue paint showing evidence of fading in three paintings by Nicolai Abildgaard, Christoffer Eckersberg and Christen Købke, highlights of the Danish collection at the Statens Museum for Kunst, have been analysed by means of XRF and Raman spectroscopy. Samples were taken from blue, purplish and greenish-blue areas of the paintings in order to investigate the colour change. In the painting by Købke, samples from both light-exposed and protected areas of the paint allowed a direct comparison between faded and non-faded colour. In the paintings by Abildgaard and Eckersberg, only non-faded areas under the rebate of the frame were sampled on the assumption that the artists used the same colours there as in the affected areas of the paint. All the areas showing fading issues were characterised by the presence of Prussian blue and lead white as main components in the blue colour. In addition, vermilion and one or more yellow pigments (Naples yellow and lead-tin yellow or Pb-Sn-Sb oxide) had been included in the mixtures by Købke and Eckersberg to produce purplish and greenish hues. The analyses carried out at the laboratories of the Centre for Art Technological Studies and Conservation (CATS) in Copenhagen confirmed the hypothesis based on visual examination of the paintings: Prussian blue was present in all the discoloured areas. In addition, the investigation highlighted the concomitant presence of lead white which, on the basis of written sources and experimental evidence together with the preparation method and size of the blue pigment, seems to be an essential factor causing the disruption on the Fe(III)–N–C–Fe(II) pathway bonding and therefore the fading of the blue colour. In some cases, as a result of the discoloration, the effect of other pigments in the blue paint mixture has been enhanced, changing the intended tonality and atmosphere of the painting.

This project, examining the use of Prussian blue in Danish paintings from the first half of the 19th century, is at a relatively early stage, but indications are that several more cases of colour change like those demonstrated above will be identified during further studies of paintings from the period. The paintings discussed in this paper are, however, good illustrations of the shortcomings of the pigment, some in a rather subtle way and others more dramatically. In a number of cases, where no parts of blue paint have been protected from the light by the frame, it is of course more difficult to assess the degree of change without sampling areas that have been protected by another paint layer, such as sky paint covered by foliage or other landscape details, and that is not always possible. Nevertheless, the examination has led to the realisation that Danish paintings of this era were sometimes created with brighter and more striking colours than the rather muted and harmonious tonality with which they are often associated by present-day viewers. In addition, the recognition of the colour changes has resulted in an art-historical reappraisal of a few individual works, as the awareness of their transformation has altered the traditional perception of these images. In the case of Christen Købke's familiar *View of Lake Sortedam* from 1838, the changes to its original appearance had altered the tonal character of the image to a degree that it resulted in an erroneous reading of the composition and the position of the painting in Købke's oeuvre.

Notes

1. One of the synonyms for the pigment, when sold in Denmark was in fact 'English blue'.

2. For details of the genesis of the individual paintings in the series, see Filtenborg 2014: 87–105.
3. Cross-sections were prepared in EasySections moulds, embedding the samples in Technovit resin and polishing using Micromesh with a granularity up to 12000. The cross-sections were photographed using a Leica DM4000 microscope.
4. Micro-Raman analyses on the painting cross-sections were carried out with a Bruker Senterra Raman microscope with a thermoelectrically cooled charged-coupled device (CCD) detector. Excitation was done with a 785 nm laser and 400 lines/mm gratings were used. The laser beam was focused with 50× and 100× Olympus objectives on the different layers of the samples with ~ 8–18 μm lateral resolution. The laser power at the sample was 1–10 mW. Accumulations between 15 and 120 scans were acquired in each spot.
5. Diary entry for 7 November 1812 in Bramsen 1947: 40.
6. Diary entry for 15 May 1839 in Villadsen 2009, 2: 825–6.
7. See note 3 above.
8. See note 4 above.
9. XRF analysis was performed using a Bruker Artax 400 X-ray fluorescence analyzer, with a Peltier cooled silicon drift detector (SDD). Analyses were performed on different areas of the painting using rhodium (Rh) target with a 1.5 mm collimator for equal live-time of 200 seconds at 50 kV and 700 μA in air.
10. See note 3 above.
11. See note 4 above.
12. XRF analyses were performed directly on different areas of the paint surface using a Bruker/S1 TITAN alloy tester energy dispersive X-ray fluorescence analyzer, with silicon drift detector (SDD). The system was equipped with a rhodium (Rh) transmission target with maximum voltage of 50 kV. Spectra were obtained with a voltage of 45 kV and 45 s of live-time acquisition, no filtration.
13. See note 4 above.
14. *Ibid.*

References

Berman, P.G. (2007) *In Another Light: Danish Painting in the Nineteenth Century*. New York: Vendome Press.

Bouvier, P.L. (1827) *Manuel des jeunes artistes et amateurs en peinture*. Paris.

Bramsen, H. (ed.) (1947) *C.W. Eckersberg i Paris: dagbog og breve 1810–13: udgivet og kommenteret af Henrik Bramsen*. Copenhagen: Thanning & Appel.

Bregnhøi, L. (2010) *Det malede rum. Materialer, teknikker og dekorationer 1790–1900*. Copenhagen: Forlaget Historismus.

Brooker, M.H., Sunder, S., Taylor, P. and Lopata, V.J. (1983) 'Infrared and Raman spectra and X-ray diffraction studies of solid lead(II) carbonates', *Canadian Journal of Chemistry* 61: 494–502.

Clark, R.J.H. and Franks, M.L. (1975) 'The resonance Raman spectrum of ultramarine blue', *Chemical Physics Letters* 34: 69–72.

de Massoul, C. (1797) *A Treatise on the Art of Painting and the Composition of Colours*. London.

Dossie, R. (1758) *The Handmaid to the Arts*. London.

Eckersberg, J. (1917) *Julie Eckersbergs optegnelser om hendes fader C.W. Eckersberg*. Copenhagen: Forening for Boghaandværk.

Filtenborg, T. (2014) *Between Formula and Freestyle: Nicolai Abildgaard and 18th-Century Painting Technique*, CATS Series of Technical Studies. London: Archetype Publications.

Frost, R.J., Martens, W.N. and Kloprogge, J.T. (2002) 'Raman spectroscopic study of cinnabar (HgS), realgar (As_4S_4), and orpiment (As_2S_3) at 298 and 77K', *Neues Jahrbuch für Mineralogie* 10: 469–80.

Gunnarsson, T. (1998) *Nordic Landscape Painting in the Nineteenth Century*. New Haven and London: Yale University Press.

Hradil, D., Grygar, T., Hradilova, J., Bezdicka, P., Grunwaldova, V., Fogas, I. and Miliani, C. (2007) 'Microanalytical identification of Pb-Sb-Sn yellow pigment in historical European paintings and its differentiation from lead tin and Naples yellows', *Journal of Cultural Heritage* 8(4): 377–86.

Jackson, D. and Monrad, K. (2010) *Christen Købke: Danish Master of Light*, exh. cat. London:

National Gallery/ Edinburgh: National Galleries of Scotland.
Kirby, J. (1993) 'Fading and colour change of Prussian blue: occurrences and early reports', *National Gallery Technical Bulletin* 14: 62–71.
Kirby, J. and Saunders, D. (2004) 'Fading and colour change of Prussian blue: methods of manufacture and the influence of extenders', *National Gallery Technical Bulletin* 25: 73–99.
Levey, M. (1989) 'Storm beneath the calm', *Art News* 88(2): 80.
Ludvigsen, L., Bagge, M. and Rask, V. (2014) 'The effect of Prussian blue on the technique of the Danish court painters Hendrick Krock and Benoit le Coffre', in H. Evans and K. Muir (eds), *Studying 18th-Century Paintings and Works of Art on Paper*, CATS Proceedings II. London: Archetype Publications, 7–15.
Merimée, J.-F.-L. (1830) *De la peinture à l'huile*. Paris.
Monrad, K. (1996) 'Silent summer evenings by Kalkbrænderi Bay and Sortedam Lake', in H.E. Nørregård-Nielsen and K. Monrad (eds), *Christen Købke 1810–1848*, exh. cat. Copenhagen: Statens Museum for Kunst, 271–7.
Monrad, K. and Conisbee, P. (1993) *The Golden Age of Danish Painting*, exh. cat. New York and Los Angeles: Hudson Hills Press in association with Los Angeles County Museum of Art.
Nørregård-Nielsen, H.E. (1996) *Christen Købke* (3 vols), vol. 2. Copenhagen: Gyldendals forlag.
Pernety, A.-J. (1757) *Dictionaire portative de peinture, sculpture et gravure*. Paris.
Samain, L., Silversmith, G., Sanyova, J., Vekemans, B., Salomon, H., Gilbert, B., Grandjean, F., Long, G.J., Hermann, R.P., Vincze, L. and Strivay, D. (2011) 'Fading of modern Prussian blue pigments in linseed oil medium', *Journal of Analytical Atomic Spectrometry* 26: 930–41.
Samain, L., Grandjean, F., Long, G.J., Martinetto, P., Bordet, P. and Strivay, D. (2013a) 'Relationship between the synthesis of Prussian blue pigments, their color, physical properties, and their behavior in paint layers', *Journal of Physical Chemistry C* 117: 9693–712.
Samain, L., Gilbert, B., Grandjean, F., Long, G.J. and Strivay, D. (2013b) 'Redox reactions in Prussian blue containing paint layers as a result of light exposure', *Journal of Analytical Atomic Spectrometry* 28: 524–35.
Sandalinas, C., Ruiz-Moreno, S., Lopez-Gil, A. and Miralles, J. (2006) 'Experimental confirmation by Raman spectroscopy of a Pb–Sn–Sb triple oxide yellow pigment in sixteenth-century Italian pottery', *Journal of Raman Spectroscopy* 37(10): 1146–53.
Schwartz, S. (1992) 'Lake Sortedam', in *Christen Købke*. New York: Timken Publishers, 49–54.
Vila, A., Monrad, K., Filtenborg, T. and Wadum, J. (2014) 'As time passed by came sunset. Christen Købke's *View of Lake Sortedam*, its genesis and colour changes', in A. Sgamelotti, B.G. Brunetti and C. Miliani (eds), *Science and Art: The Painted Surface*. Cambridge: Royal Society of Chemistry, 354–72.
Villadsen, V. (ed.) (2009) *C.W. Eckersbergs Dagbøger 1810–1853*, 2 vols. Copenhagen: Nyt Nordisk Forlag Arnold Busck.
Wadum, J., Monrad, K. and Scharff, M. (2010) 'Aspects of Christen Købke's painting technique: from drawing via oil sketch to the final painting', in M. Spring with H. Howard, J. Kirby, J. Padfield, D. Peggie, A. Roy and A. Stephenson-Wright (eds), *Studying Old Master Paintings: Technology and Practice*. London: Archetype Publications, 216–21.
Wivel, M. (1993) *Christen Købke*. Copenhagen: Edition Bløndal.

Authors' addresses

- Troels Filtenborg, Statens Museum for Kunst, Copenhagen, Denmark (Troels.Filtenborg@smk.dk)
- David Buti, CATS, Statens Museum for Kunst, Copenhagen, Denmark (David.Buti@smk.dk)
- Anna Vila, CATS, Statens Museum for Kunst, Copenhagen, Denmark (Anna.Vila@smk.dk)
- Jørgen Wadum, CATS, Statens Museum for Kunst, Copenhagen, Denmark (J.Wadum@smk.dk)

'Paintings fade like flowers': colour change in paintings by Vincent van Gogh

Ella Hendriks

Introduction

The Dutch painter Vincent van Gogh (1853–1890) was a colourist in heart and soul: 'one can't do without it; one must make use of it', he once wrote about colour.[1] He became famous in particular for the bold expressive palette of his later French works, yet early on, when still in the Netherlands, he had already defended the idea that the artist had a right to adopt a personal use of colour and should start with the colours on his palette, rather than slavishly copy the colours in nature. 'COLOUR EXPRESSES SOMETHING IN ITSELF', he wrote in capital letters.[2] He believed that painters should be 'free to seek colours that form a *whole* and are related to one another', and furthermore, that this relation between colours that looked beautiful and were 'therefore right', as he put it, not only applied to individual works, but extended to the colour relationship between works that formed a series, a well-known example being the series of *Sunflower* paintings he made to decorate the Yellow House in Arles. Van Gogh's careful attention to colour carried right through to the finishing touches in his paintings, adding signatures and painted borders to his compositions for a note of contrasting colour (Figures 1 and 2), and advising appropriate colours for the frames.[3]

Van Gogh's ideas on tone and colour can primarily be traced back to the theory of complementary colours, which became known through the work and pronouncements of the French painter, Eugène Delacroix (1798–1863). He had read about Delacroix's colour theories in 1884 in Charles Blanc's *Grammaire des arts du dessin: architecture, sculpture, peinture*, a book in which Blanc further developed the laws of simultaneous colour contrast originally put forth by Delacroix and by the French chemist, Michel Eugène Chevreul (1786–1889). Van Gogh considered these laws as on a par with the importance of Newton's explanation of the laws of gravity or Stephenson's laws of steam.[4]

Blanc's explanation read as follows:

> If we combine two of the primary colours, yellow and blue, for instance, to compose a binary colour, *green*, this binary colour will reach its maximum of intensity if we place it near its complement – *red.* So, if we combine yellow and red to form *orange*, this binary colour will be heightened by the neighbourhood of *blue.* Finally, if we combine red and blue to form *violet*, this colour will be heightened by the immediate neighbourhood of *yellow.* Reciprocally, the red placed beside the green will seem redder; the orange will heighten the blue, and the violet the yellow. It is the reciprocal heightening

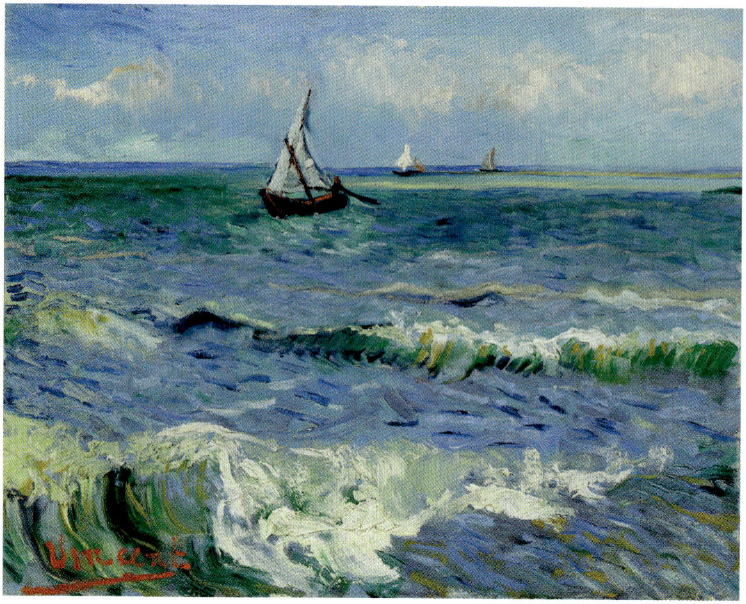

Figure 1 Vincent van Gogh, *Seascape near Les Saintes-Maries-de-la-Mer*, June 1888, oil on canvas, 64.3 × 50.5 cm, Van Gogh Museum, Amsterdam. In a letter dated 13 August1888, Van Gogh explained how back in the studio he added an 'outrageous signature' to this seascape painted on the beach in Saintes-Maries-de-la-Mer because he 'wanted a red note in the green'.

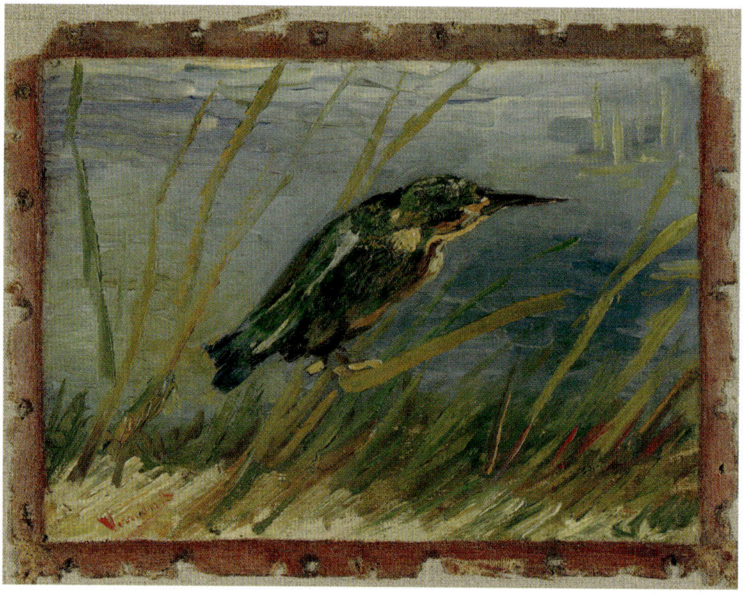

Figure 2 Vincent van Gogh, *Kingfisher by the Waterside*, Paris, July–August 1887, oil on canvas, 26.6 × 19.1 cm, Van Gogh Museum, Amsterdam. Van Gogh painted a red border around the picture that was to be exhibited unframed.

> of complementary colours in juxtaposition that M. Chevreul called '*The law of simultaneous contrast of colours*' (Blanc 1889: 145–69; see also van Dijk 2013: 216–17) (Figure 3).

In his paintings Van Gogh made extensive use of bright simultaneous colour contrasts for emotive effect, inspired by the example of Delacroix. For example, he praised the contrast of the lemon yellow halo within the dramatic violet, dark blue and blood-red patch of figures, set off by the emerald sea, in Delacroix's painting of *Christ Sleeping During the Storm* (*c.*1853) (Figure 4).[5] In Paris he also became inspired by the rich colour usage of Adolphe Monticelli (1824–1886), a recently deceased painter from Marseilles, later claiming: 'as far as I know there is no colourist who comes so straight and directly from Delacroix'.[6] Like Delacroix, it is thought that Van Gogh played with skeins of wool to try out arrangements of colour for his paintings. A set of woollen balls in a small Chinese lacquerwork chest has come down to us along with other paraphernalia from his Paris studio and shows close parallels with the colour schemes of his pictures (Figure 5).

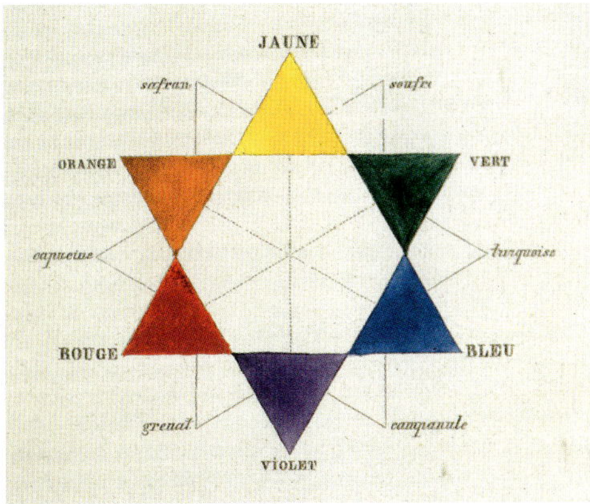

Figure 3 Charles Blanc, colour wheel (reproduced in *Grammaire des arts du dessin: architecture, sculpture, peinture*, Paris 1867).

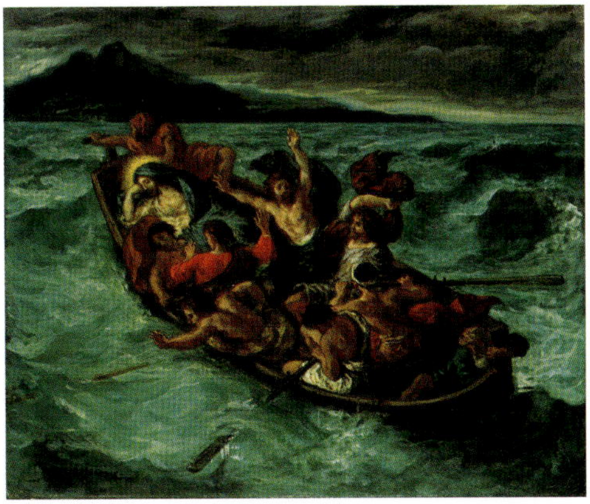

Figure 4 Eugène Delacroix, *Christ on the Sea of Gennesaret*, *c.*1853, oil on canvas, 50.8 × 61 cm, The Metropolitan Museum of Art, New York, Bequest of Mrs. H. O. Havemeyer, 1929, The H.O. Havemeyer Collection.

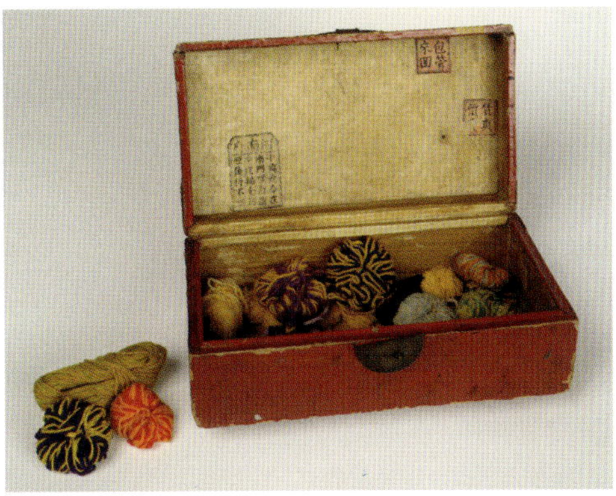

Figure 5 Red lacquered box containing balls of wool, 19th century, Van Gogh Museum, Amsterdam.

Van Gogh on colour change

Given the primary importance of colour to Van Gogh it is hardly surprising that his letters express a concern for the adverse effects of ageing and consequent colour change. 'Paintings fade like flowers', he wrote following a visit to the municipal museum in Montpellier, adding that even some of Delacroix's had suffered.[7] Throughout his life as an artist, Van Gogh was dependent on his brother Theo for money to buy painting materials and the artist's letters provide ample evidence for his quest to balance price with quality when selecting his purchases. Writing about the expensive but durable pigment cobalt blue, for example, he acknowledged that 'the quality of the colour may not be everything in a painting, but all the same that's what gives it vitality'.[8] He consulted paint manufacturers for their advice on sound colours, such as 'the greens that persist',[9] and was well aware of the poor reputation of certain pigments such as Prussian blue, chrome yellow and the so-called 'geranium lake' prepared from eosin dyestuff (*Lacque geranium*), ranking them among the unstable colours made fashionable by the Impressionists.[10] Nonetheless, enticed by the unparalleled effects he could achieve with these colours, he continued to use them until his death in July 1890.[11] Surviving paint orders record that between April 1888 and June 1890, when working in Arles, Saint-Rémy and Auvers-sur-Oise, he ordered at least 38 tubes of the vivid but especially fugitive geranium lake colour, with far-reaching consequences for his paintings as they are seen today.[12] Anticipating that these unstable colours would change over time, he suggested 'all the more reason boldly to use them too raw, time will only soften them too much'.[13]

Besides pigment change, Van Gogh also took into account the yellowing of the oil binding medium. To reduce the oil content in the paint that would otherwise yellow over time, he requested that his suppliers grind his pigments more coarsely, providing fresher colours that would last longer,[14] and wrote that he would return freshly ground tubes of colour if they appeared too greasy.[15] He once even suggested shaving down the extremities of impasto with a razor blade, thereby removing the surplus oil that accumulates at the surface and would otherwise turn yellow.[16] For the same reason he recommended washing the surface of some paintings with water, or water with a little spirits of wine.[17] Rather than reflecting the idiosyncratic practices of a painter who was largely self-taught, in fact these measures demonstrate Van Gogh's broader knowledge of artists' tricks of the trade and working methods, as laid down in artists' handbooks, for example.[18] While these preventive measures have undoubtedly helped to some extent, nowadays we may conjecture that the degree to which the colour schemes of his paintings (and drawings) have altered over time presumably lies beyond that which he could reasonably have imagined.

Colour and conservation

Given the key importance of colour for the artist, it is an aspect of primary concern behind efforts at the Van Gogh Museum to conserve and display Van Gogh's legacy of works in an appropriate way. This paper focuses on the question of pigment deterioration brought about by light exposure, yet it would be telling only half the story not to also briefly mention efforts to address consequences of colour change caused by past restoration treatments. Most significantly, in the period 1926–1933, virtually the entire family collection of Van Gogh's paintings (which now forms the basis of the Van Gogh Museum collection) was subjected to a systematic campaign of wax-resin lining and varnishing treatments performed by the Dutch conservator, J.C. Traas (1898–1984).[19] Whereas Van Gogh considered it inevitable that sooner or later his quickly painted studies would have to be lined,[20] he would have had the traditional French method of glue-paste lining in mind (once described to him in detail by his colleague Paul Gauguin),[21] rather than the Dutch method of infusing with wax-resin that is known to cause darkening of absorbent layers of ground and paint. In general, we now choose to leave the old wax-resin linings in place, but research over the past two decades has greatly expanded our knowledge of the priming materials used by Van Gogh and in conjunction with experimental reconstructions has improved our ability to assess the degree to which lining has caused passages of exposed canvas and ground to darken in individual works (Hendriks 2011b; Nieder *et al.* 2011). Most treatments at the Van Gogh Museum involve the painstaking removal of the old, usually dammar resin, varnish layers

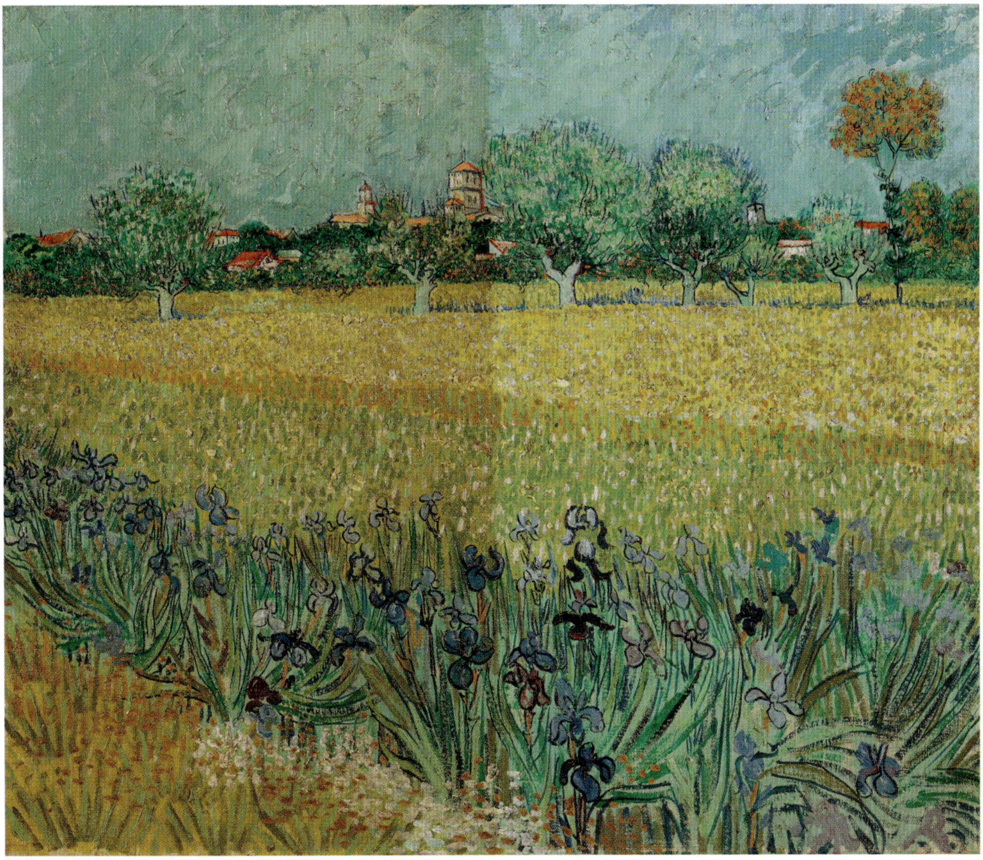

Figure 6 Vincent van Gogh, *Field with Irises: Arles in the background*, May 1888, oil on canvas, 54.3 × 64.8 cm, Van Gogh Museum, Amsterdam. The picture is shown halfway through removal of the old and discoloured varnish layer in 2014.

applied by Traas that have yellowed over time and run counter to the artist's general preference for a matt, unvarnished surface in his French-period paintings (Figure 6).[22]

Chrome yellow and red lake studies

A second focus at the Van Gogh Museum and the main topic of this paper is preventive conservation aimed at slowing down the degradation of unstable colours. To date, our studies have tended to focus on two classes of light-sensitive pigments that are especially important for the colour changes observed in Van Gogh's paintings: on the one hand chrome yellow pigments that may darken, and on the other red lake pigments that may fade, both through light exposure.

The poor reputation of chrome yellow was well known among painters, including Van Gogh, who read about the danger of 'chrome yellow changing more than gold and turning green with time' in Théophile Silvestre's memoirs on Delacroix.[23] Recent scientific investigations in collaboration with colleagues at the universities of Antwerp and Perugia have shown that it is the paler, lead chromate-lead sulphate variety of the pigment that is susceptible to darkening upon light exposure through reduction of the Cr(VI) ion to Cr(III).[24] This transformation takes place in a very thin region (typically 2 µm or less in paint cross-sections)

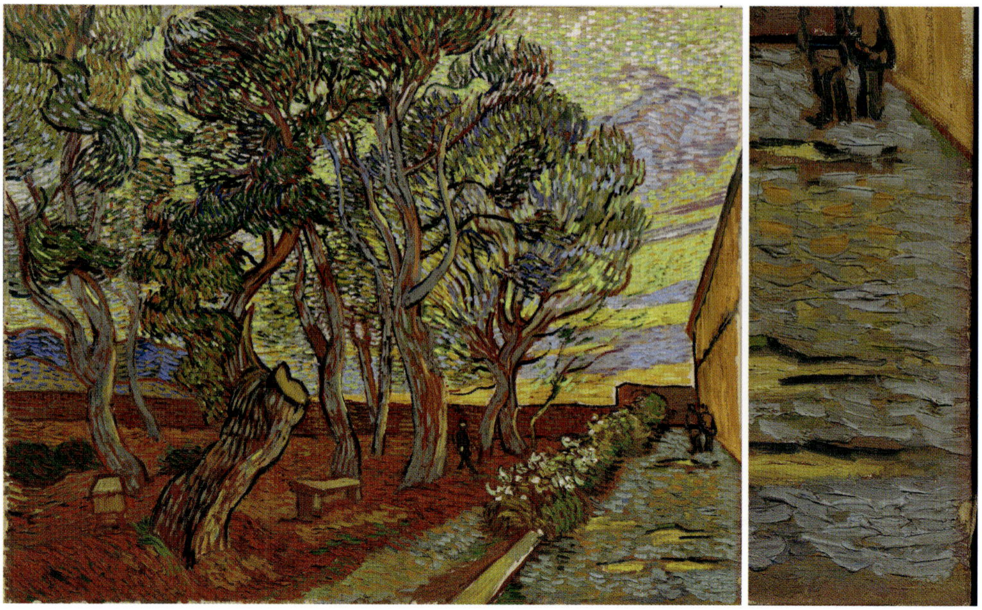

Figure 7 Vincent van Gogh, *Garden of the Asylum*, November 1889, oil on canvas, 72 × 91 cm, Van Gogh Museum, Amsterdam. A strip of brighter purple colour is preserved down the right edge of the picture under the frame where the light-sensitive pigment, brazilwood lake, has been protected from light.

at the surface of the paint, so that a careful distinction must be made between darkening of the paint surface due to reduction of the chromium ion, as opposed to the visual effects of darkening caused by saturation of the paint surface by a smooth skin of oil or accumulated grime and discoloured varnish (Roy and Hendriks forthcoming).

Case-study publications signalling colour change due to the fading of fugitive red lake pigments in Van Gogh's paintings began to appear from the turn of the 21st century (see, for example, Cadorin 2001 and Rioux 1999), and increasingly the scale of this phenomenon has emerged. At the Van Gogh Museum in Amsterdam and the Kröller-Müller Museum in Otterlo, comprehensive and systematic campaigns of technical examination conducted for new collection catalogues of Van Gogh's paintings revealed repeated evidence for the fading of red lakes, sometimes in the form of a brighter margin of colour preserved where the edges of a painting have been protected from light (Figure 7).[25] These cumulative observations have led to growing awareness of the fact that such colour shifts due to the fading of red lake pigments are commonplace in Van Gogh's French oeuvre and indeed affect virtually every work. Observation combined with analytical studies performed by scientists at the Cultural Heritage Agency of the Netherlands has enabled us to chart the different types of red lake pigment used by Van Gogh throughout his career, and to recognise characteristic types of degradation affecting his works (van Bommel *et al.* 2005).[26] In the past, these colour changes in Van Gogh's paintings have largely gone unnoticed and hence have not been taken into account in art-historical analyses of the style and content of individual works (Figure 8).

Furthermore, within the framework of the 'De Mayerne Programme' supported by the Netherlands Organization for Scientific research (NWO), accelerated ageing studies of model reconstructions were undertaken to improve our understanding of the relative light stabilities of the different types of red

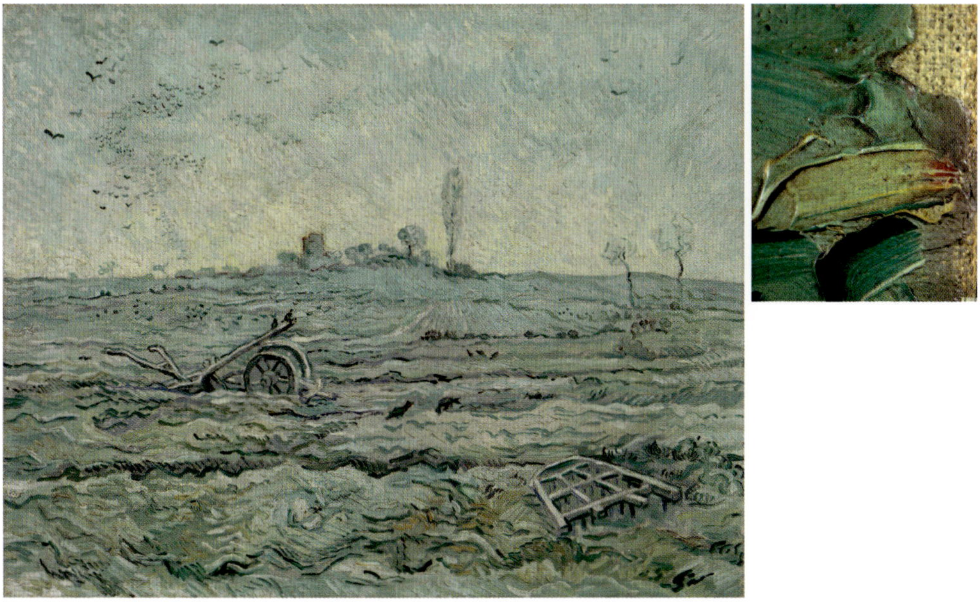

Figure 8 Vincent van Gogh, *Snow-covered Field with a Plough (after Millet)*, January 1890, oil on canvas, 72.3 × 91.6 cm, Van Gogh Museum, Amsterdam. The cold monochrome blues and greens of this winter landscape have been said to reflect the intense loneliness that the painter experienced when out alone in the countryside. However, the edges of the painting show remnants of unfaded cochineal lake (on an aluminium and calcium-containing substrate), revealing that its original colour scheme was warmer than it is now.

lake pigment and the rate at which they fade. In these test reconstructions, the most fugitive lakes faded in less than 500 hours of light ageing, which is equivalent to less than 10 years of exposure to museum lighting conditions.[27] It is therefore possible and even likely that these redwood and eosin-containing paints may have faded within or soon after Van Gogh's short lifetime, especially in view of the uncontrolled lighting conditions that then prevailed.

Reversing the effects of colour change in a digital visualisation

With this increased awareness of colour change due to fading of red lakes came a growing desire to be able to visualise its effect. After all, talking about things is not the same as seeing it for oneself. Although we realised from the start that a digital visualisation reversing the effects of discoloration would not result in a precise image of how the paintings looked when they were executed, with all their detailed nuances, it was felt that it can be a useful tool to help us come closer to imagining the general colour relationships and effects that the artist intended.

In 2009–10, a project to conserve and restore the version of Van Gogh's *Bedroom* painting, now in Amsterdam, prompted the first widescale effort to make a simulation of its unfaded colours. In the course of technical examination and treatment, multiple forms of evidence were found for colour change having taken place in this painting, in which colour played a key role for the artist. The painting of his bedroom on the first floor of the Yellow House in Arles was meant to convey the sense of calm he experienced there after 10 days of such hard work that strain to his eyes had forced him to take a few days off. Besides

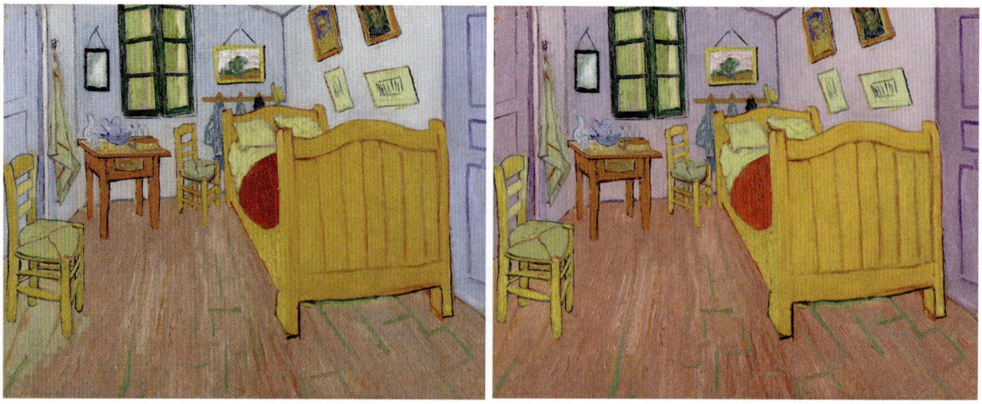

Figure 9 Vincent van Gogh, *The Bedroom*, Arles, October 1888, oil on canvas, 72.4 x 91.3 cm, Van Gogh Museum, Amsterdam. An image of the painting as it is now (left) is shown beside a digital visualisation of its colour scheme before fading of red lake pigments occurred (right).

the emptiness of the interior and the sturdy and simple character of the furniture, it was above all colour that would give the picture its impact. 'The colour has to do the job', as he put it, or rather a simplification of colour, for it had to create a suggestion of calmness or sleep, bringing peace of mind and giving his imagination a rest.

Van Gogh describes the colour scheme of the painting in detail in two letters, yet comparison of the colours of the painting now with those described by the artist reveals that the colour scheme of the painting has changed significantly, departing from what was intended. The changes, due mainly to the fading of geranium and cochineal lakes, are described in detail elsewhere, as well as the highly interdisciplinary process of creating a digital visualisation that attempts to reverse the effects of colour change in the *Bedroom* painting (Hendriks *et al.* 2010; Fiedler *et al.* 2016).[28] Only the result is shown here, illustrated beside an image of the painting as it looks now (after the 2010 restoration treatment) so that the two can be compared directly (Figure 9).

There are many differences but one striking feature is the way that once the false sense of space given by the current light blue background (which acts like a sky) is eliminated, the cramped space of the bedroom interior is evoked more effectively. The original violet colour of the walls conveys the effect of dim light in the closed interior, against which the chinks of yellow sunlight stand out more clearly. Importantly, the recovery of a balanced complementary colour scheme (the colours no longer compete with each other) seems more in keeping with Van Gogh's intention that the painting should be suggestive 'of rest' or 'of sleep'.[29] The overall effect is closer to that of a Japanese print 'coloured in flat, plain tints', as intended by Van Gogh.[30]

Sharing insights

Until 2010, no one had ever questioned the way the *Bedroom* painting looked with its blue walls and blue doors, and cold pink floor. In that year, when the painting was returned to the gallery, we wanted to include the visualisation of the original colour scheme in a display presenting the outcome of the conservation research and treatment, but were uncertain how the public would react. We decided to show it in the same room as the newly restored painting, but at some distance from it, so that the viewer could look back and forth from one to the other, or focus on the painting alone if desired.

In 2014 the digital simulation of the *Bedroom* was incorporated in an app in the series *Touch van Gogh – What Paintings can Reveal*, allowing the user to explore the painting layer by layer through the interactive display.[31] A computer installation with the app now forms a permanent feature beside the painting. Previous evaluation of using digital media in conservation displays at the museum had shown that the app device needed to be close enough to the object for its connection to be clear, and while it was initially feared that this would detract from the painting, it has been found that interactives in the gallery actually prolong visitor engagement and encourage the viewer to repeatedly return to the painting and look again in response to new information they offer or questions they raise (de Vet and van Kregten 2014).

Meanwhile, since 2010, showing digital simulations of objects in an earlier state of preservation has become increasingly commonplace. In the case of Van Gogh, we may even speak of a trend as digital simulations form the focus of two major exhibitions held in 2015 and 2016 respectively.[32] In a broader sense, sharing these images with the general public has triggered mainstream awareness of the fact that this type of museum object, of an age between that of modern and 'Old Master' paintings, is still subject to ongoing change. It could even be argued that in turn this awareness has paved the way for understanding and acceptance of the need to reduce cumulative light exposure by reducing light levels in the museum, as this paper goes on to discuss.

The future?

But what about the future? Besides helping to envisage the original appearance of the *Bedroom* painting, as part of this project we examined the potential of using digital simulations to look forward in time and anticipate future effects of colour change, working together with risk management expert Agnes Brokerhof at the Cultural Heritage Agency of the Netherlands (RCE). The continued fading of the red lakes is expected to have the main impact.[33] Whereas cochineal lake in the blue background and doors has already faded to the extent that its colour can no longer be seen, with continued light exposure the fading of geranium lake will result in the floor becoming even cooler in tone, while the red lines on the bed will eventually fade and disappear.

Using spectral data acquired from the accelerated ageing of red lake reconstructions, print simulations were made to visualise the colours of the painting in progressive states of deterioration as a result of continued light exposure (Figure 10). In random sequence, a series of different sets of two prints was shown to members of staff and management at the museum, the left print always less faded than the right one.[34] Each viewer was asked to define the perceived level of change in terms of a percentage loss in value, and to state in how long a period of time this change should be allowed to take place.[35]

Based on the prints shown, staff and management at the Van Gogh Museum opted for the lowest level of light damage (equivalent to 10% loss of value in fading of the red floor) considered acceptable over an approximate 30-year period. This translated into a light budget of 4 million lux-hours (4 Mlxh) spread over 30 years, roughly just one-third of that which we had been spending! This outcome prompted a re-evaluation of the standard maximum illumination level of 150 lux in the galleries, causing us to more than halve it to a range of 50–70 lux.

While it is acknowledged that accelerated ageing of model paint systems and computer simulations will not provide a result that precisely mirrors the effects of natural ageing in paintings, we believe that the procedure has worked well in our museum. Questioning staff and management based on the digital simulations has triggered awareness of light damage occurring to the collection, as well as new ways

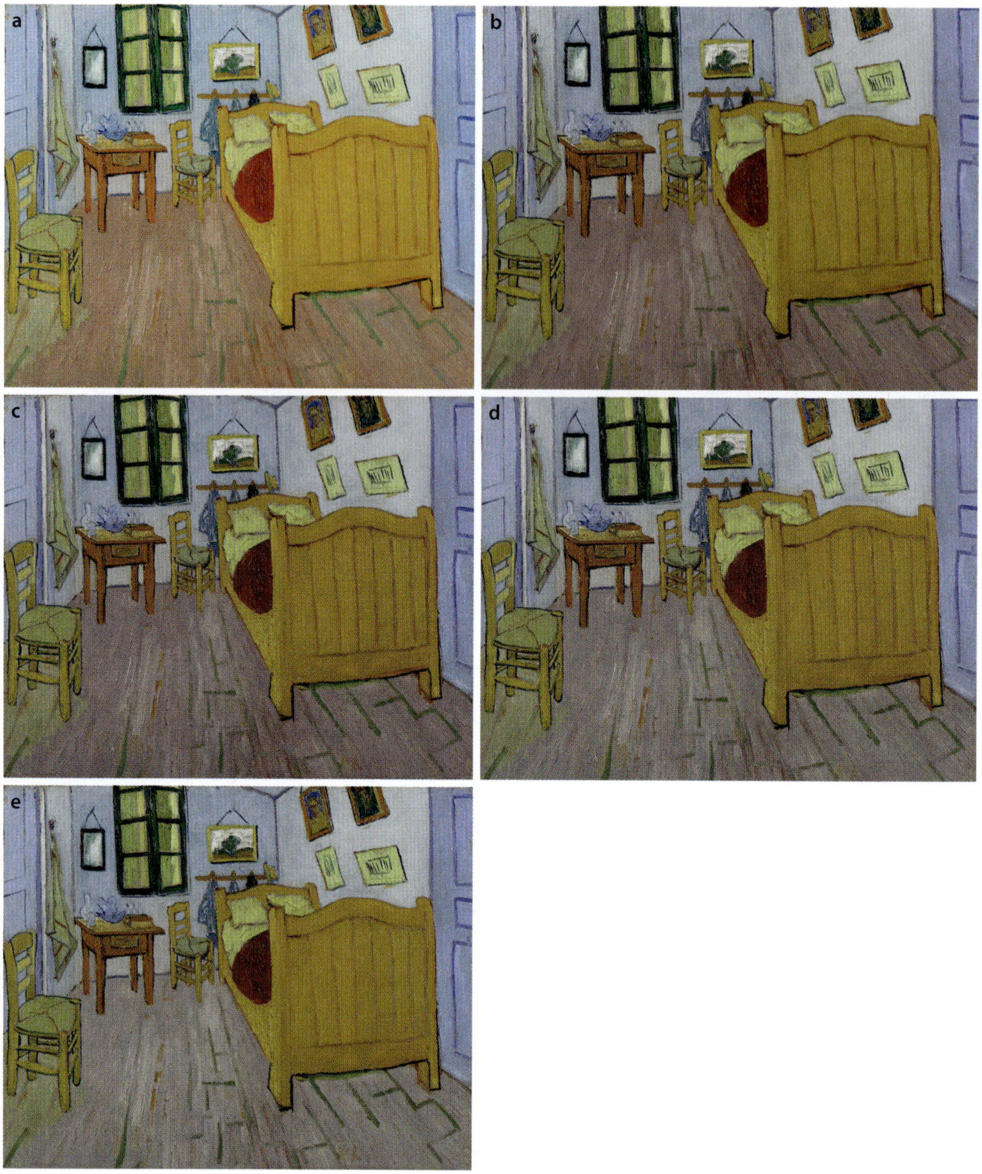

Figure 10 Digital simulations predicting continued colour change in *The Bedroom* painting due to the varied light exposure. The painting is shown as it looks now (a) and as it will probably look after exposure to 4 million lux-hours (4 Mlxh) (b), 8 Mlxh (c), 15 Mlxh (d), and 30 Mlxh (e).

of thinking about its consequences. Indeed this could be seen as its main purpose, without immediately providing 'the right answer', which will have to develop as we go through a continual process of assessing and reassessing the changes that take place as a consequence of our decisions and actions. In concrete terms, since there was a need to formulate a basis for our lighting policy and decide on an allowable dosage, the visualisations provided a tangible basis for discussion and for making informed decisions. In general, since light damage is irreversible, we err on the side of caution and choose to act now rather than when it is too late. The chosen guidelines can always be tightened, or relaxed, as new insights become available through continuing research.

Notes

1. Letter 537, to Theo van Gogh, Nuenen, on or about Wednesday, 28 October 1885. The artist's letters contain numerous references to colour as a theme of central importance. The edition of Van Gogh's letters quoted in this paper is the comprehensive searchable digital resource first made available in 2009 by Leo Jansen, Hans Luijten and Nienke Bakker: http://vangoghletters.org/vg/.
2. Letter 537, cited in note 1 above.
3. Regarding the framing of his works Van Gogh once wrote, 'I can't finish off except in a frame' (letter 699). He painted them in various colours or used white ones to enhance the colours within the painting, as well as plain wooden ones (oak, chestnut, deal and walnut) with a distinctive tone. On white frames, see letters 705 and 834, and for wooden frames, letters 683, 685, 699 and 778; see also Jooren 2013: 299–305.
4. Letter 538.
5. Letter 632, to Emile Bernard. Arles, Tuesday, 26 June 1888.
6. Letter 853.
7. Letter 765.
8. Letter 549.
9. Letter 547.
10. Letter 595. See also letter 538 on the poor stability of the chromate yellow pigments and letter 687 on Prussian blue.
11. On the late French palette of Van Gogh, see Geldof *et al.* 2013a.
12. The paint orders are included in letters 593, 654, 674, 687, 710, 758, 863 and 877; see also Geldof *et al.* 2013b.
13. Letter 595.
14. Letters 676 and 687.
15. Letter 593.
16. Letter 538.
17. Letters 775 and 853.
18. Scholarly research presented to a broader public in the recent exhibition *Van Gogh at Work* held at Van Gogh Museum, Amsterdam from 1 May 2013 to 12 January 2014, has laid to rest this popular notion of the artist as an isolated and temperamental genius who set to work without any prior thought, simply using whatever materials were at hand. See the various contributions in Vellekoop *et al.* 2013a; see also Vellekoop *et al.* 2013b.
19. For the past conservation and restoration of paintings in the Van Gogh family collection, see Hendriks 2011a.
20. Letter 800.
21. Letter 734.
22. On this topic see Jooren 2013: 291–4.
23. Silvestre 1863: letter 526 n. 4.
24. In the period 2011–2014 the results of this research were presented in a 5-part serial publication in *Analytical Chemistry*. See articles by Monico *et al.* 2011, 2015.
25. The margins of preserved colour are due to the fact that the paint has been shielded from light, either by the frame, or by brown paper tape applied around the edges of the paintings as a customary measure after lining treatment by the restorer J.C. Traas. Since often we know when Traas lined the paintings in question, this indicates the time frame in which the observed colour change must have continued to take place.
26. On geranium lake see Geldof *et al.* 2013b and Burnstock *et al.* 2005. We are grateful to Shell Nederland, 'Partner in Science' to the Van Gogh Museum, for offering facilities and expertise for SEM-EDS analysis of red lake samples.
27. The red lake pigments were made in a workshop coordinated by Jo Kirby in 2005; see Kirby 2013. These pigments were used to create paints in a workshop coordinated by Leslie Carlyle in 2005; see Carlyle 2013. For the results of artificial ageing tests see Burnstock *et al.* 2005 and van den Berg *et al.* 2006.
28. The team for the digital reconstruction of the original colours of the *Bedroom* painting included art historians (Louis van Tilborgh and Teio Meedendorp), a conservator (Ella Hendriks), artist (Monica Rotgans), imaging

specialist (Maurice Tromp) and conservation scientist (Muriel Geldof), working together with colour scientist, Roy S. Berns of the Munsell Color Science Laboratory, Rochester University, New York.

29. Letter 705.
30. *Ibid.*
31. *Touch Van Gogh – What Paintings can Reveal:* an innovative series of apps that are free to download for iPad and Android tablets (IOS 6 or higher) from Apple and Google Play app stores.
32. *Irises and Roses*, held at the Metropolitan Museum of Art, New York, 12 May–16 August 2015. See: www.metmuseum.org/metmedia/video/collections/ep/van-gogh-irises-and-roses. *Van Gogh's Bedrooms*, 14 February–8 May 2016, The Art Institute of Chicago; see Fiedler *et al.* 2016.
33. The chrome yellow used for the bed has been identified as the light-stable variety of the pigment (the monoclinic, pure lead chromate) and hence is not expected to be vulnerable to darkening upon light exposure.
34. We are grateful to Maurice Tromp, imaging specialist at the Van Gogh Museum, for making the required prints.
35. This ongoing research has been expanded to incorporate other audience groups including students from the University of Amsterdam conservation training programme and members of the museum public, and forms the topic of a forthcoming publication. A similar experimental approach is described in Brokerhof *et al.* 2008.

References

Blanc, C. (1889) *The Grammar of Painting and Engraving*, K.N. Doggett (trans.). Chicago.

Brokerhof, A.W., Reusl, M., Mackinnon, F., Ligterink, F., Neevel, H., Fekrsanati, F. and Scott, G. (2008) 'Optimum access at minimum risk: the dilemma of displaying Japanese woodblock prints', in D. Saunders (ed), *Conservation and Access. Contributions to the London Congress 15–19 September 2008*. London: International Institute for Conservation, 82–7.

Burnstock, A., van den Berg, K.J., Carlyle, L., Clarke, M., Hendriks, E., Kirby, J. and Lanfear, I. (2005) 'A comparison of the fading and surface deterioration of red lake pigments in six paintings by Vincent van Gogh with artificially aged paint reconstructions', in J. Bridgland (ed.), *Preprints of the ICOM-CC 14th Triennial Meeting*, vol. 1. London: James and James, 459–66.

Cadorin, P. (2001) 'Colour fading in Van Gogh and Gauguin', in C. Peres, M. Hoyle and L. van Tilborgh (eds), *A Closer Look: Technical and Art-Historical Studies on Works by Van Gogh and Gauguin*, Cahier Vincent 3, 2nd edn. Zwolle: Waanders, 12–19.

Carlyle, L. (2013) 'Historically accurate oil paint reconstructions', in M. Clarke, J.H. Townsend and A. Stijnman (eds), *Art of the Past: Sources and Reconstructions*. London: Archetype Publications, 53–9.

de Vet, M. and van Kregten, J. (2014) 'Touch Van Gogh and be touched: how new media are transforming the way we present complex research', in N. Proctor and R. Cherry (eds), *Museums and the Web 2014*. Silver Spring, MD: Museums and the Web.

Fiedler, I., Hendriks, E., Meedendorp, T., Menu, M. and Salvant, J. (2016) 'Van Gogh's *Bedrooms*: materials, intention and evolution', in G. Groom *et al.*, *Van Gogh's Bedrooms*, exh. cat., 14 February–8 May 2016, The Art Institute of Chicago, 68–103.

Geldof, M., Megens, L. and Salvant, J. (2013a) 'Van Gogh's palette in Arles, Saint-Rémy and Auvers-sur-Oise', in M. Vellekoop, E. Hendriks, L. Jansen, M. Geldof and A. de Tagle (eds), *Van Gogh's Studio Practice*. Brussels: Van Gogh Museum/Mercatorfonds (distributed by Yale University Press), 238–55.

Geldof, M., de Keijzer, M., van Bommel, M., Pilz, K., Salvant, J., van Keulen, H. and Megens, L. (2013b) 'Van Gogh's geranium lake', in M. Vellekoop, E. Hendriks, L. Jansen, M. Geldof and A. de Tagle (eds), *Van Gogh's Studio Practice*. Brussels: Van Gogh Museum/Mercatorfonds (distributed by Yale University Press), 268–89.

Hendriks, E. (2011a) 'Treatment history of the collection', in E. Hendriks and L. van Tilborgh with

M. van Eikema Hommes and M. Hageman, *Vincent van Gogh, Paintings 2, Antwerp and Paris (1885-1888)*, Van Gogh Museum. Zwolle: Waanders, 28–36.

Hendriks, E. (2011b) 'Van Gogh's working practice: a technical study', in Ella Hendriks and Louis van Tilborgh with the assistance of Margriet van Eikema Hommes and Monique Hageman, *Vincent van Gogh, Paintings 2, Antwerp and Paris (1885-1888)*, Van Gogh Museum. Zwolle: Waanders, 97–112.

Hendriks, E., Jansen, L., Salvant, J. *et al.* (2010) 'A comparative study of Vincent van Gogh's *Bedroom* series', in M. Spring with H. Howard, J. Kirby, J. Padfield, D. Peggie, A. Roy and A. Stephenson-Wright (eds), *Studying Old Master Paintings: Technology and Practice*. London: Archetype Publications, 237–43.

Jooren, M. (2013) 'Van Gogh's finishing touches: varnish, signatures, frames and painted borders', in M. Vellekoop, E. Hendriks, L. Jansen, M. Geldof and A. de Tagle (eds), *Van Gogh's Studio Practice*. Brussels: Van Gogh Museum/ Mercatorfonds (distributed by Yale University Press), 290–305.

Kirby, J. (2013) 'Reconstructions of French 19th-century red lake pigments used at the time of Vincent van Gogh', in M. Clarke, J.H. Townsend and A. Stijnman (eds), *Art of the Past: Sources and Reconstructions*. London: Archetype Publications, 69–77.

Monico, L., van der Snickt, G., Janssens, K., de Nolf, W., Miliani, C., Verbeeck, J., Tian, H., Dik, J., Radepont, M., Hendriks, E., Geldof, M. and Cotte, M. (2011) 'The degradation process of lead chromate in paintings by Vincent van Gogh studied by means of synchrotron X-ray spectromicroscopy and related methods. Part 2: Original paint layer samples', *Analytical Chemistry* 83(4): 1224–31.

Monico, L., Janssens, K., Hendriks, E., Vanmeert, F., van der Snickt, G., Cotte, M., Falkenberg, G., Brunetti, G. and Miliani, C. (2015) 'Evidence for degradation of the chrome yellows in Van Gogh's *Sunflowers*: a study by non-invasive in situ methods and synchrotron radiation-based X-ray techniques', *Angewandte Chemie International Edition*, DOI: 10.1002/anie.201505840.

Nieder, E., Hendriks, E. and Burnstock, A. (2011) 'Colour change in sample reconstructions of Vincent van Gogh's grounds due to wax-resin lining', *Studies in Conservation* 56(2): 94–103.

Rioux, J.-P. (1999) 'The discoloration of pinks and purples in Van Gogh's paintings from Auvers', in A. Distel and S.A. Stein (eds), *Cézanne to Van Gogh: The Collection of Doctor Gachet*. New York: Metropoltan Museum of Art, 104–14.

Roy, A. and Hendriks, E. (forthcoming) 'Van Gogh's *Sunflowers* in London and Amsterdam', *National Gallery Technical Bulletin* 35.

Silvestre, T. (1863) *Eugène Delacroix. Documents Nouveaux*. Paris.

van Bommel, M., Geldof, M. and Hendriks, E. (2005) 'The analysis of red organic pigments in paintings by Vincent van Gogh', *Art Matters: Netherlands Technical Studies in Art* 3: 129–53.

van den Berg, K.J., Burnstock, A., Carlyle, L., Clarke, M., Hendriks, E., Hoppenbrouwers, R., Kirby, J. and Lanfear, I. (2006) 'Fading of red lake paints after Vincent Van Gogh: an interdisciplinary study involving three De Mayerne projects', in J.J. Boon and E.S.B. Ferreira (eds), *Reporting Highlights of the De Mayerne Programme*. The Hague: NWO, 89–96.

van Dijk, M. (2013) 'Van Gogh and the laws of colour: an introduction', in M. Vellekoop, E. Hendriks, L. Jansen, M. Geldof and A. de Tagle (eds), *Van Gogh's Studio Practice*. Brussels: Van Gogh Museum/Mercatorfonds (distributed by Yale University Press), 216–25.

Vellekoop, M., Hendriks, E., Jansen, L., Geldof, M. and de Tagle, A. (eds) (2013a) *Van Gogh's Studio Practice*. Brussels: Van Gogh Museum/ Mercatorfonds (distributed by Yale University Press).

Vellekoop, M. with Bakker, N., van Dijk, M., Geldof, M., Hendriks, E. and Reissland, B. (2013b) *Van Gogh aan het werk/Van Gogh at work*, exh. cat. Brussels: Van Gogh Museum/Mercatorfonds.

Author's address

Ella Hendriks, Van Gogh Museum, Amsterdam, The Netherlands (e.hendriks@vangoghmuseum.nl)

Colour change and modern paints: an overview

Bronwyn Ormsby and Judith Lee

Introduction

As a distinct area of investigation, colour change in relation to paints used from the second half of the 20th century onwards remains relatively unexplored. This may be due in part to the implementation of quality control standards which have ensured that colour change is minimal.[1] Nonetheless, several largely interrelated phenomena can influence how colour is perceived in modern and contemporary paintings. The types of changes noted include: yellowing; darkening; changes in gloss; whitening; greying; increased/decreased transparency; glossy medium-rich surfaces and drips, in addition to fading and other changes in pigment hue.

Exploring the range of physical and chemical changes behind these effects can be particularly complex, involving aspects such as: the erosion and/or embrittlement of medium (or coating) layers; the formation of surface crusts/layers from the migration and/or degradation of paint constituents; the formation of surface layers due to atmospheric interactions with paint surfaces; decreased physical coherence due to the loss of lower molecular weight materials and changes to molecular weight; changes in pigment crystal size, structures and composition; changes in paint volume and density, as well as deposited soiling and conservation treatment.

The aesthetic quality of unvarnished modern and contemporary painting surfaces can also play a particularly important role in the interpretation, integrity, authenticity and even value of a work of art. Therefore continuing efforts to understand the range of changes that can take place is important for interpretation and conservation treatment design, as well as wider collections care. This paper presents an overview of known factors that can influence colour change in modern paints by drawing on literature, artist interviews, discussions with artists' paint manufacturers, and observations from past and current collaborative research projects at Tate, with an emphasis on the two paints that have dominated the artists' paint market since the 1960s: artists' oil and acrylic dispersion paints.[2]

Quality control standards for modern paints

One of the key advantages of paints made since the 1970s over their historic counterparts is the presence of a group of international and national standards that have been developed as guidelines for permanence, quality control and health, pertaining to paints, paint constituents and associated evaluation methods. The American Society for Testing Materials International (ASTM) and the British Standards Institution (BSI) are the most relevant bodies for the regulation of artists' paints.[3] The first ASTM standard, *Paint and Related Coatings, Materials, and Applications*, was listed as D1

Table 1 Current ASTM International standards for the evaluation of paints, pigments, colour change, hiding power and lightfastness.

ASTM Standard	Scope
D4303-10	For testing the lightfastness of colorants used in artists' materials. Four test methods to accelerate the effects of long-term indoor illumination on artists' materials
D2244	Practice for Calculation of Colour Tolerances and Colour Differences from Instrumentally Measured Colour Coordinates
D4302 - 14	Specification for Artists' Oil, Resin-Oil, and Alkyd Paints – establishes requirements for composition, physical properties, performance, and labelling of artists' oil, resin-oil, and alkyd paints. This specification covers pigments, vehicles, and additives. Lightfastness of the paints shall be tested to meet requirements
D5067	Specification for Artists' Watercolour Paints
D5098 – 05a	Specification for Artists' Acrylic Dispersion Paints – establishes requirements for composition, physical properties, performance, and labelling of artists' acrylic emulsion paints. Requirements are included for pigment identification, lightfastness, bleeding, consistency, and drying time
D5398	Practice for the Visual Determination of the Lightfastness of Art Materials by the User, Blue Wool comparisons.
D5383	Practice for the Visual Determination of the Lightfastness of Art Materials by Art Technologists

in 1902 (ASTM International 1998). Although there are now standard specifications for acrylic emulsion, watercolour, gouache, oil, resin-oil and alkyd paints (Table 1), which represent the 'current consensus between producers, consumers and other interested parties of minimally acceptable performance characteristics and other criteria', they were not developed for artists' paints in particular. The current ASTM *Standard Test Methods for Lightfastness of Colorants used in Artists' Materials*, D4303 (ASTM 2010), is however specific to artists' paints, and includes guidelines for exposure to exterior conditions under glass, artificial daylight fluorescent lamps and xenon-arc lamps.

A range of ASTM artists' paint pigment ratings was created by Henry Levison who built on a long history of lightfastness and permanence testing in Europe (Levison 1976). ASTM ratings are now included on individual paint tube labels as:

- ASTM I – excellent
- ASTM II – very good or good
- ASTM III – fair (and not sufficiently lightfast to be used in artists' paints)

The scale proceeds further to IV (poor) and V (very poor), but these ratings are not used on artists' paints. To meet the ASTM standard specifications for use in artists' paints, samples must exhibit a mean colour change of less than eight colour difference units under the specified evaluation conditions, designed to predict performance in a typical household or museum environment (Gavett 2012). British Blue Wool standards, which were initially created for the evaluation of textiles,[4] are also used to assess the permanence and lightfastness of painted artworks via microfadometry.[5] The Blue Wool rating scale for pigment lightfastness ranges from 8 (absolutely lightfast) to lower numbers which are progressively less so. As an example, Golden Artist Colors compare a Blue Wool 6 rating to an ASTM rating of II (very good) and a Blue Wool 7 rating to ASTM I (excellent) (Golden Artist Colors 1995).

Artists' paint manufacturer evaluations

Artists' paint manufacturers carry out their own evaluations of lightfastness and/or permanence. For example, Schmincke's lightfastness rating scheme involves an accredited five-star system based on Blue Wool ratings, where colours are exposed to sunlight on a rooftop at a 45° slope facing south, and the process is repeated whenever formulations change (Schminke 2013). Chroma also employ Blue Wool evaluations for pigments and paints as well as samples covering a range of tints as 'pale tints were never tested by ASTM and it is in the palest tints that unreliable colours fade most' (Chroma Australia 2016). As described above, Golden Artist Colors use a combination of modified ASTM and Blue Wool standard comparisons. Golden state that products with ASTM ratings of less than II are not suitable for artists' products, those with ratings of II (very good) are expected to remain unchanged (with respect to colour) for approximately 100 years, and ratings of III and beyond should not be used in works of art intended to last (Golden Artist Colors 1995). Sands (2015) describes the process of submitting new data to ASTM by Golden Artist Colors as follows:

> the only way to add a color to a list is to put it through all the testing specified by ASTM, including both prolonged outdoor and accelerated indoor exposures. Once completed, those results are submitted to the committee and if approved the color can be included on the list and its ASTM rating placed on a label. From start to finish this process usually takes several years.

Winsor & Newton (W&N) define permanence for an artists' colour as 'its durability when laid with a brush on paper or canvas, graded appropriately and displayed under a glass frame in a dry room freely exposed to ordinary daylight and an ordinary town atmosphere', and that the evaluations are intended to provide a measure of the likelihood of fading. Accelerated tests for lightfastness and binder stability are combined with results from pigment manufacturers' testing. Lightfastness is assessed by reducing colours (not watercolours) to a level of 40% reflectance through the addition of titanium white, and samples are tested in both sunlight and artificially accelerated conditions. All W&N paint tube labels include an ASTM lightfastness rating and a W&N permanence rating, defined as: AA – Extremely Permanent; A – Permanent; B – Moderately Durable and C – Fugitive (Winsor & Newton 2016a).[6]

Colour change and synthetic organic pigments

The vast increase in numbers and types of synthetic organic pigments throughout the 20th century is well documented (Christie 2002; Herbst and Hunger 2004; Learner 2005; Quillen-Lomax and Learner 2006; Keijzer 2014), and some guides to lightfastness ratings are available (Levison 1976; Berrie 2007; Myers 2013). Jones *et al.* (2005: 10) note that 'most pigments used in contemporary artist paints are selected from a list of about 120 pigments that are established to be highly light fast in acrylics, and in oils'. Perhaps unsurprisingly there are relatively few publications on colour change pertaining to synthetic organic pigments within conservation, although investigations into pigments used in the 1950s, 60s and 70s may prove useful.

Synthetic organic pigments are helpfully classified through the Colour Index (CI),[7] a database founded in 1924 by the Society of Dyers and Colourists. The current pigment database exceeds 500 pigments, with a few thousand colorants in total. The key to CI classifications (Sands 2007) are as follows:

- Colourant type (P = pigment; D = dye; S = solvent dye)
- Hue (e.g. R = red, Y = yellow etc)
- Index number (pigments are assigned a sequential number when added)
- Constitution number (five-digit number based on chemical structure)
- Common names (general use names for a pigment)

One well-known example of the problematic use of synthetic organic pigments in modern paintings includes Mark Rothko's use of the synthetic organic pigment lithol red (PR49) in his Seagram mural paintings. This pigment's deterioration has been determined as a primary cause of the substantial colour change noted in several of his works (Carlyle *et al.* 2008; Stenger *et al.* 2010;[8] Townsend *et al.* 2014). In a study of the history of PR49,

Standeven (2008) stated that the 'primary drawback of lithol red is its poor lightfastness' and that it was primarily used in student grade colours. The fading behaviour of PR49 is however also dependent on the paint medium type(s) used, the lithol red salt form (i.e. sodium, calcium, barium, strontium) and exposure to high relative humidity (RH) (Stenger *et al.* 2010; Kennedy *et al.* 2012). During a recent treatment of Rothko's painting *Untitled: Black on Maroon*, 1958 (Tate T01170), test samples prepared with PR49:1 (barium form) and ultramarine blue pigment in rabbit skin glue faded to a greater extent when accelerated light ageing was combined with cycling RH (55–85% RH) rather than the same conditions with a stable 55% RH (Barker and Ormsby 2015).

Despite a few pioneering papers exploring the fading behaviour of synthetic organic colorants (Whitmore and Cass 1989; Berrie and Quillen-Lomax 1997), the primary focus of scientific research has involved the analytical identification of pigments in paints and works of art (Learner 2005; Quillen-Lomax 2010a, 2010b; Quillen-Lomax *et al.* 2007; Scherrer *et al.* 2009). More recently, the research of Gherlardi *et al.* (2015) involved a multi-instrumental assessment of the lightfastness of 18 synthetic organic pigments without any binder (PB1, PO5, PO16, PO46, PR3, PR48:1, PR49:2, PR53:1, PR57:1, PR83, PR112, PR170, PV1, PV27, PV39, PY3, PY12, PY100), some of which are still used in paints. Samples were artificially aged using ASTM D4303-10 protocols (some ultraviolet radiation (UV), 55% RH, raised temperatures), after being mixed with isopropanol and spread onto glass slides. Colour measurement data showed that there was no correlation between the pigment chemical class and degree of colour change, and that the ASTM lightfastness ratings of pigments in specific binders matched with the ΔE^* measured for the powders for only a few pigments (e.g. PB1, PO5, PR83, PR112, PR170). For the other pigments included, the lightfastness performance proved better than the ASTM ratings when pure, which reinforces the role of binders in enhancing photo-induced fading. Where greater levels of colour change occurred, chemical analysis facilitated a correlation with structural changes, such as loss of auxochrome groups, and other chemical changes such as yellowing (increased chromophores and auxochromes), largely induced by oxidative processes.

Regarding issues pertaining to conservation treatment, some pigments are known to have poor resistance to organic solvents. Studies carried out by the authors showed that acrylic paints with azo yellow pigment (PY3) cleaned with mineral spirits produced a maximum ΔE^{94} of 0.1, with some pigment transfer. Blumenroth *et al.* (2014) investigated the solubility of 23 synthetic organic pigments dispersed in oil paint to six different solvents, however colour change data were not published. Defeyt and Strivay (2014) investigated six polymorphs of phthalocyanine blue (PB15) pigments (PB15:3 is most common to artists' paint formulations). They explored the potential for colour change resulting from induced crystallisation defects in PB15:0 when exposed to aromatic solvents, which potentially causes colour shifts towards green but this did not occur under the conditions explored. They did note, however, that the paints remained highly solvent sensitive and occasionally water sensitive, regardless of the polymorph present.

Colour change and acrylic dispersion paints

Early acrylic paints

Paint manufacturers tend to have a positive view on the ageing properties of acrylic paints. Levison apparently 'shared his enthusiasm for the reduction of yellowing in acrylic paints … compared to traditional oil paints' (Golden 2010a). When discussing the durability of acrylic paints, Jones *et al.* (2005: 10) state that:

Table 2 Synthetic organic pigments used in current W&N Professional Acrylic paints (Winsor & Newton 2016b).

Hue	W&N Professional Acrylic Colours: synthetic organic pigment CI type and index numbers
Yellow	PY3, PY83, PY110, PY129, PY139. PY150
Red	PR112, PR122, PR149, PR170, PR177, PR179, PR206, PR209, PR254, PR255
Violet	PV19, PV23, PV29
Green	PG7, PG36
Blue	PB15:1, PB 15:3, PB16, PB60
Orange	PO49, PO73

> formulators of artist paints used commercially available latexes that had been designed for other uses, primarily for exterior trade sales paints. It was tacitly assumed that if an acrylic is durable enough to use outdoors, it can be expected to last for a long time indoors ... *[and that this]* assumption has been validated by experience.

However, they also draw attention to some of the problems, stating that early acrylic 'artist paints had several drawbacks relative to oil paints. For example, they were less transparent, they had less gloss, they were more susceptible to dirt pickup, the colours changed as they dried, and the paintings were harder to clean.' Relating to the early days of W&N acrylic paint production, Ian Garrett[9] (formerly of W&N) recalled that natural burnt sienna and raw sienna pigments tended to cause acrylic dispersion media to 'liver up', i.e. become sticky and form complexes in paint tubes due to the presence of soluble salts in these naturally sourced pigments. Garrett also mentioned that other issues encountered with paint surfaces today were probably related to excess surfactant, choice of surfactant and/or an excess of preservatives.

On his use of acrylic paints, British artist John Hoyland reported that some of the early Spectrum paints he had went hard in the jars (presumably due to the use of hard tap water in paint formulations) and that one of his purple colours turned 'white' (Ormsby *et al.* 2014: 537). Some of Hoyland's paintings have been identified as having fine layers of migrated surfactant on paint surfaces, and one of the mixed green passages from *25.4.69* (Tate, T01129) proved to be unusually water-soluble, presumably due to the use of excessive levels of surfactant in early paint formulations.

The first W&N acrylic paint lines took advantage of the newer, more lightfast pigments except for genuine burnt and raw sienna, in addition to pigment pastes which were made for industrial paints, containing finely ground pigments, added humectants and preservatives. Ian Garrett confirmed that W&N used single pigments where possible for their professional lines and that lake pigments were no longer needed as PR170 (naphthol red) and PV23 (dioxazine violet) were now used. Garrett also noted that some of the early synthetic organic pigments such as PY1 (azo Hansa yellow) and PR3 (toluidine red) were later replaced with more permanent options. Table 2 lists the current range of synthetic organic pigments used in Winsor & Newton's Professional Acrylic colours (Winsor & Newton 2016b) with associated permanence ratings of AA or A. These paints also have minimal wet to dry colour shift, which presumably has been achieved through the use of microemulsion binders (Ormsby *et al.* 2012).

Yellowing and discoloration

With respect to colour change, earlier studies into acrylic media noted that the causes of discoloration included migration of components from canvas/wood supports and polymer cross-linking reactions, some of which reversed when (re)exposed to visible light (Jablonksi *et al.* 2004). More recently, unpigmented acrylic dispersion media have been reported to discolour to some extent by Smith (2007: 238) and Golden Artist Colors (Golden 2010b). Golden reported on a range of gel media where the 'more matte, and thickly applied gel coats ... age to a more opaque and more amber to brown appearance'. They also expect 'the thick gloss gels to also have … a slight loss of clarity'. At Tate, naturally aged clear gel medium samples from a Liquitex board (Figure 1) are all notably soiled and yellowed. At this point, the exact constituents and reactions responsible for these types of colour changes have not been elucidated.

Discoloured migrated surfactant layers are regularly referred to within the house paint industry,[10] often accompanied by advice about rinsing the layers off, and statements on how removal will not affect paint durability. However the general discoloration of fine, migrated surfactant layers on works of art has not been reported. Another area where yellowing can be a significant concern is the application of acrylic solution and acrylic dispersion copolymer varnishes which may discolour similarly to gel media. Examples of discoloured acrylic-resin varnished paintings have begun to appear, including one work with a grey and soiled acrylic solution varnish (Ormsby *et al.* 2011), another with a discoloured acrylic dispersion varnish, and a third with an artist-applied, discoloured styrenated-acrylic dispersion varnish. These coatings pose significant ethical and conservation treatment challenges, particularly when the coatings are artist-applied and the underlying paint films are acrylic dispersion.

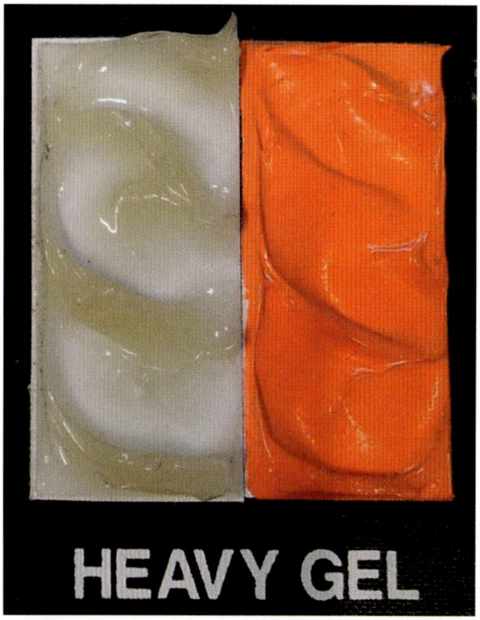

Figure 1 Detail of a Liquitex advertisement board showing discoloured and yellowed High Viscosity Heavy Gel medium and paint sample (© Tate 2016).

Accelerated ageing studies

One key study (Learner *et al.* 2002) explored the thermal ageing colour stability of a range of white artists' paints where the acrylic paints fared very well in comparison to oil and alkyd equivalents. A later study (Ormsby *et al.* 2007) reported that a range of titanium white (PW6) and azo yellow (PY3) artists' acrylic dispersion paints exposed to accelerated light ageing (no UV) did not result in significant colour change ($\Delta E^{94} < 0.4$), however thermal ageing resulted in the visible yellowing of some titanium white paints, with a maximum ΔE^{94} of ~2.5. More recently, Izzo *et al.* (2014) carried out a study comparing commercial oil, acrylic and polyvinyl acetate artists' paints with three pigment types (chrome green (PG17), Hansa yellow (PY74) and titanium white) using three different accelerated ageing regimes: ozone, high RH and thermal. Colour change (yellowing) was unsurprisingly greatest for the thermally aged oil paints and some change (less than ~3 ΔE units) was noted for the acrylic and PVA samples. They also concluded that ozone exposure did not appear to affect the acrylic and PVA paints (and possibly helped to stabilise the oil paints), and that humidity affected the acrylic and PVA paints more than the oil paints.

When UV radiation was included in accelerated ageing, acrylic paints predictably showed more change. Doménech-Carbó *et al.* (2011) explored the daylight exposure (some UV) of acrylic dispersion and PVA paints and noted that more significant colour change depended on the pigment and binding medium present, with zinc oxide in an acrylic dispersion medium registering the least colour shift. In general, both chroma and hue were

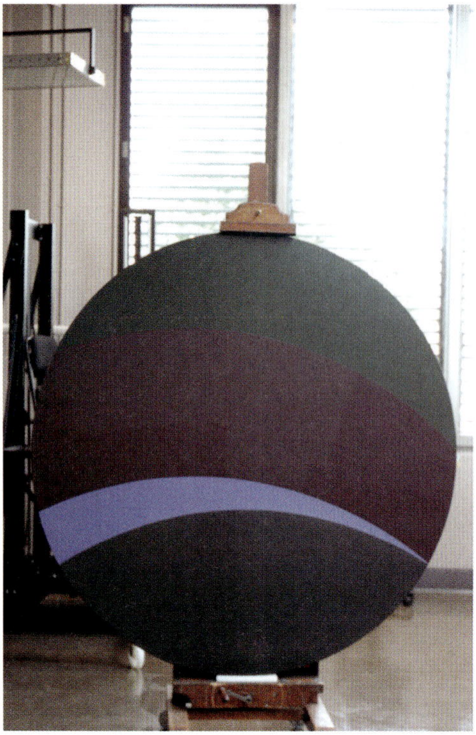

Figure 2 Overall image of Alexander Liberman's painting *Andromeda T00650* during surface cleaning. The darker, more saturated areas have been cleaned (Tate 2007 © Alexander Liberman Trust).

affected negatively and shifts in lightness (L*) were associated with changes in the binding medium. Paints containing inorganic pigments tended to exhibit the least change, and for the organic pigments evaluated, the phthalocyanine group registered slightly less colour shift. It was also noted that anatase TiO_2 accelerated the degradation of the polymer, which led to colour change.

In another study (Pintus and Schreiner 2011), acrylic dispersion paints were aged under UV conditions. It was found that inorganic pigments increased the reactivity of the acrylic binding media and caused the formation of unspecified aldehydes, lactones and acidic oxidation products. Another paper by the same group (Pintus *et al.* 2015) explored the UV photo-oxidative stability of a range of paints including oil-modified alkyd, acrylic dispersion and polyvinyl acetate (PVA) containing eight different inorganic pigments. It was noted that the pigments caused an increase in sensitivity to UV light and that the largest colour shifts were associated with alkyd paints. They also observed that the most pronounced colour change was associated with cadmium yellow pigments and anatase titanium white.

Paint surfaces

In their review, Jablonksi *et al.* (2004) noted that acrylic dispersion paint surfaces can suffer from haziness, largely caused by either the migration of intrinsic, water-soluble surfactant materials to paint surfaces and/or a milky effect that can appear during the drying process due to the retention of moisture within the film. The migration of surfactant, particularly in combination with deposited soiling, can form hazy, whitish layers on acrylic paintings that can have a significant effect on paint saturation and gloss. Recent studies suggest that the presence/absence of migrated surfactant on painting surfaces may be related (to some extent) to pigment content, hence some pigments may play a role in photo-induced surfactant degradation at paint surfaces (Ormsby *et al.* 2008). Affected surfaces may appear streaky, an effect that is often more visible on darker paints. Recently, the streaky, hazy surface of an acrylic dispersion and PVA painting was found to contain migrated surfactant material, gypsum (which had also migrated from the paint film) and deposited soiling (Hillary and Ormsby 2014). In another recent example, a hazy and discoloured layer on an acrylic painting contained nicotine in addition to migrated surfactant materials (Fremout *et al.* 2014).

Several studies have been carried out into the effects of surface cleaning treatments on acrylic paints and so far colour change has been attributed to the removal of soiling and migrated surfactant, resulting in a significant recovery of paint saturation and a general increase in surface gloss (see Figure 2)

(Ormsby *et al.* 2013). Due to the inherent softness and porosity of acrylic paint films however, there is a risk that soiling may become permanently embedded, leading in some cases to permanent colour change. Further research is required to explore the effect of environment, paint quality, brand and soiling type on the embedding of deposited matter into acrylic dispersion paints. Golden Artist Colors (Golden 2010a) suggest that matt acrylic paintings may absorb 'more pollution discoloration that will be difficult to remove', however they also state that 'for paintings constructed with more color than gel or other medium ... the colors will remain quite crisp as long as a regular maintenance of dust removal or varnish application is considered'.

Colour change and modern oil paints

Detailing the molecular changes that occur during the drying, alteration and degradation processes of oil paints is beyond the scope of this paper. However, there are colour change phenomena associated with modern oil-containing paints that pose complex challenges (Burnstock and van den Berg 2014). The range of issues includes: yellowing and darkening; physical deterioration; light-scattering surface layers resulting from the formation of metal soap crusts; fatty acid efflorescence; possible oxalate layers; the formation of water-soluble medium 'skins,' and in extreme cases the dripping of yellowed medium material. It is no surprise that these phenomena can contribute to visual change making works of art with these issues problematic to conserve. Research into the effective surface cleaning of vulnerable modern oil paintings has begun to appear,[11] however much remains to be done, in particular towards developing appropriate, low-risk conservation treatment strategies.

Efforts have been made in recent years to assess a range of degradation phenomena based on pigments and pigment–media interactions that can lead to oil paint colour change, such as changes to synthetic ultramarine blue (Cato *et al.* 2016), conversion of cadmium pigments to other cadmium salt forms (Pouyet *et al.* 2015), and a range of surface and bulk paint changes induced by the formation of metal carboxylate products (Helwig *et al.* 2014; Osmond 2014). For example, in one project (Zhang 2013), titanium white (TiO_2) pigments were mixed with five colorants in linseed oil and aged using UV. Results showed that the colour (fading, darkening), gloss (decrease) and film roughness (increase) had changed after ageing, and that the most change had occurred where an uncoated anatase TiO_2 pigment was used. It was also noted that the green earth pigment was not affected, the blackening of vermilion increased, that Prussian blue degraded faster and that a synthetic organic pigment Hansa yellow (possibly PY3, PY74) also degraded more in the presence of TiO_2.

Ian Garrett recalled that W&N experienced early issues with anatase-based TiO_2 oil paints with added Prussian blue, leading to colour change within a few weeks, and that the use of the anatase pigment resulted in complaints regarding 'sticky colours.' He also reported that natural iron oxides could be affected by shade problems depending on the source, and that zinc chromate (zinc yellow, PY36) was not lightfast (as well as being toxic) and was replaced with synthetic organic pigments such as PG7 (phthalocyanine green), PY3 (azo or Hansa yellow) and then PY74 (arilyde yellow) in the early 1970s. Some W&N oil colours – for example, Winsor Orange (lake pigment, used before PO13 and PO72) – also reacted with flake white and lost colour. Other problematic colours included Winsor Violets (containing PV1, PV2), geranium colours (lakes), organic yellows, chromate colours such as lead zinc chromate, as well as lemon yellow (barium chromate) and arsenical lakes (mauve shades). It is interesting to note that both cadmium (Mass *et al.* 2013) and chrome colours (Monico *et al.* 2016) are currently being investigated for deterioration in oil paints.

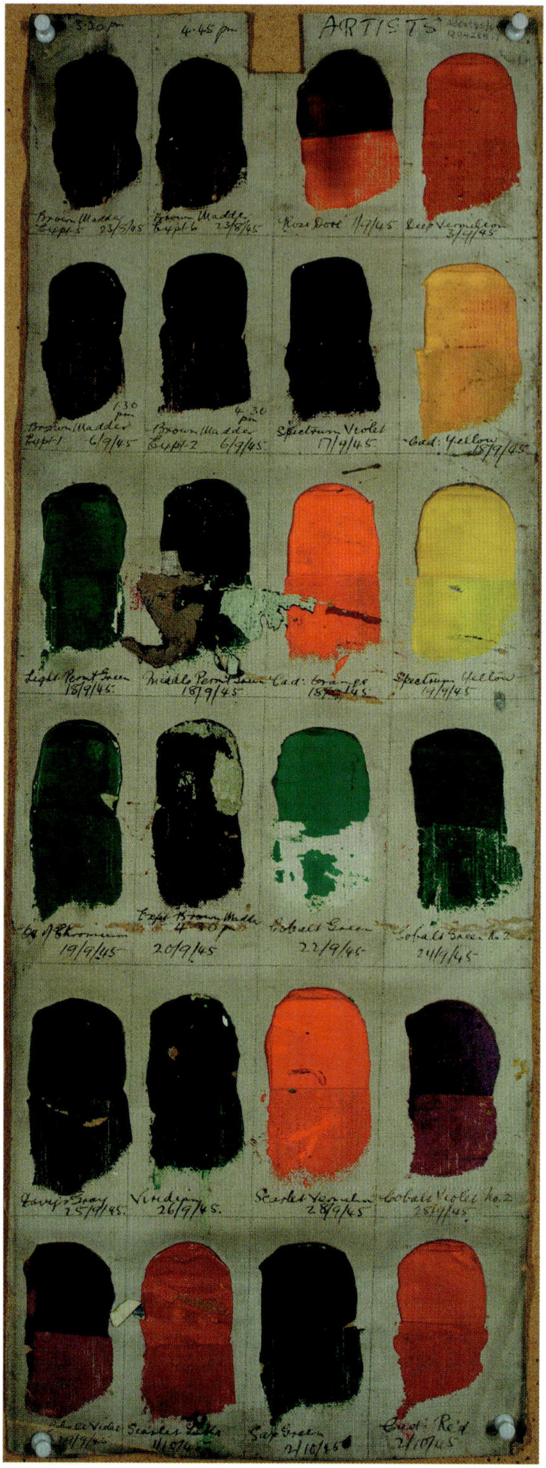

Figure 3 Example of a W&N artists' oil paint swatch, *AOC1945_01_PC* (© Tate 2014).

Exploring colour change in Winsor & Newton oil paint swatches (1945–2003)

Tate holds 268 'Artists' Oil' colour swatches produced by Winsor & Newton as part of its quality control procedures. Each paint sample (Figure 3) represents a single application of colour over prepared ground, or paper support, with two thicknesses. These swatches were stored in environmentally uncontrolled conditions for many years, and have been the subject of an ongoing collaborative study with the Courtauld Institute of Art, London, exploring several phenomena that are also found on modern oil paintings. The types of colour change noted on the swatches include: darkening and yellowing due to dark storage; medium separation and darkening (e.g. French ultramarine); the formation and yellowing of medium skins; the formation of dark and light surface crusts and fatty acid efflorescence. The following sections detail some of these observations.

Yellowing and surface dirt

A combination of surface dirt and yellowing/darkening caused by dark storage had the most significant impact on the colour of the W&N swatches, particularly those dating from 1945 to 1966. Yellowing was most evident for white paints of the later swatches, for example, lead white (1970s–2003), titanium white (1966–2003) and zinc white (1970–2003). The effect of surface dirt on the appearance of paintings is well documented elsewhere (Eastaugh 1990). In some W&N samples, dirt and particulate or fibrous debris had become embedded within yellowed medium skins, such as Aureolin, Lemon Yellow and Cadmium Yellow paint films dating from the 1940s.

The discoloration of oil paints as a result of dark storage, and subsequent bleaching upon exposure to light, is well known (Carlyle 2001; Carlyle *et al.* 2002; Levison 1985; Rakoff *et al.* 1979). Yellowing occurs most rapidly within the first few days of dark storage (van den Berg *et al.* 2002), with subsequent yellowing being less extensive than that caused by the initial dark storage. Oils with a high degree of unsaturation such as linseed oil are often said to yellow the most, however after a significant period of ageing the degree of yellowing for both linseed and poppy seed oil can be similar (Mallégol *et al.* 2001). The white W&N swatches made with combinations of poppy and safflower oils had still yellowed significantly. The yellowing commonly associated with linseed oil that occurs when the drying is not complete may be due to impurities (Mallégol *et al.* 2001). Other factors influencing the degree of yellowing include the proportion of oil in the paint, oil processing (e.g. heat treatment) and drier concentration (Carlyle *et al.* 2002; Levison 1985). Pigments also influence the yellowing and fluorescence of linseed oil (Rie 1982). Interestingly, the intensity of the UV fluorescence of the W&N swatches appeared to correlate with the apparent thickness of (yellowed) medium skins which had developed on many of the paints.

Fatty acid efflorescence

As frequently seen on unvarnished modern oil and alkyd-based works of art (Aufdermarsh 1988), many of the W&N swatches exhibited fatty acid efflorescence. This largely consists of palmitic and stearic acids migrating from the oil medium (Singer *et al.* 1995), resulting in a white, light-scattering surface which can have a dramatic effect on colour. It has been shown that fatty acid efflorescence is liable to return after removal within periods of under 10 years (Burnstock *et al.* 1993; Perry 1990).

The mechanisms for the migration of saturated fatty acids in oil-based paintings are not fully understood. One possibility is the idea of syneresis, whereby physical 'squeezing pressures' caused by the cross-linking and contraction of the paint film can physically push out mobile components. Alternatively, as oxidation reactions cause the oil medium to become more polar, the hydrophobic long-chain fatty acids become increasingly incompatible with

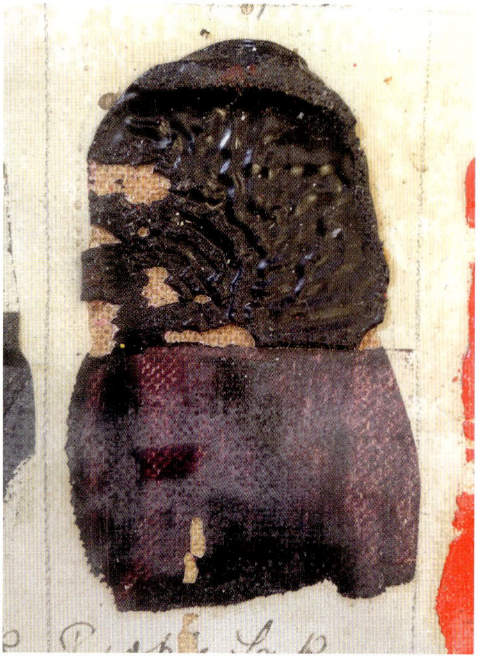

Figure 4 Fatty acid efflorescence on the surface of a Purple Lake W&N oil swatch dated 1948 (© Tate 2015).

the paint film thereby promoting 'phase' separation and migration to the surface. Stabilisation of free fatty acids, by metal ions capable of forming metal soaps, is believed to inhibit fatty acid efflorescence (van den Berg *et al.* 1999).

The presence of oil-rich grounds, pigments with a high oil absorption such as alizarin (Koller and Burmester 1990) and slow-drying paint are also associated with fatty acid efflorescence (Singer *et al.* 1995). Other factors include: pigment particle size; additives including metal stearates, free fatty acids or wax; the use of slow-drying and semi-drying oils such as walnut, safflower and soya oils (Caldwell 2002); and the absorbency and permeability of the support (Saunders and Kirby 2004). The W&N swatches most affected by fatty acid efflorescence appeared to be Alizarin Crimson, Burnt Umber, Lamp Black, Transparent Golden Ochre, Winsor Red, Winsor Green, Winsor Lemon and Winsor Violet. These colours contain either organic colorants or pigments with relatively high oil absorption, where there are fewer stabilising interaction opportunities between free fatty acids and metal ions within the paint film.

Of the W&N swatches examined, fatty acid efflorescence mostly affected older paints (Figure 4), pre-1960s, possibly because of their longer exposure to fluctuating environmental conditions. Fluctuating temperature and RH conditions and/or exposure to high humidity are known to exacerbate the fatty acid efflorescence seen in paintings (Åckerlund 2012; Hinde *et al.* 2011). In addition to the type of pigment present and the presence of light (Keune *et al.* 2008), high RH is known to promote hydrolysis of acylglycerides and oxidative cleavage of fatty acid side chains (as well as the light-induced colour change of certain pigments), which would result in an increased concentration of free (mobile) fatty acids. However, given that the degree of hydrolysis of aged paint films has been shown to reach up to 94% (van den Berg *et al.* 1999), further increases in the degree of hydrolysis may be relatively insignificant, and would not promote significant increases in fatty acid efflorescence. Therefore, other possibilities such as humidity gradients, which promote the diffusion of moisture through paint films, may cause swelling and hence facilitate the migration of the free fatty acids already present within the paint film (Michalski 1991).

Metal soaps

Metal soaps may be present in a paint film as additives, for example, aluminium (Tumosa 2001) and magnesium stearates.[12] Metal soaps may also form *in situ* as a result of the reaction between metal ions and free fatty acids. While metal soaps are generally desirable in terms of their stabilisation of aged oil paint films (Boon *et al.* 1997), they can also form insoluble surface crusts and/or a white haze that is intimately bound to the paint surface, and which can disfigure the image (Sawicka *et al.* 2014). Surface crusts are believed to be caused

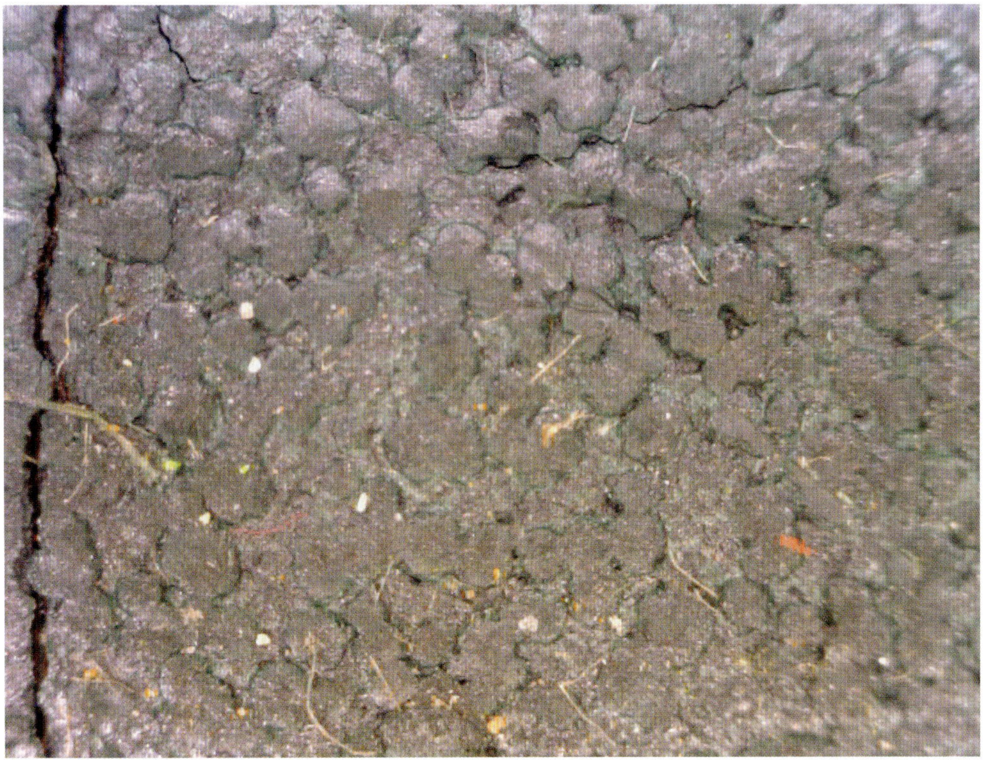

Figure 5 Detail of a surface crust on a Viridian W&N oil swatch from 1947 (© Tate 2015).

by the migration and subsequent remineralisation of metal soaps, and may be composed of mixtures of salts (e.g. sulphates, carbonates or oxalates), the exact composition of which will vary depending on the composition of the paint film and the anionic composition of the atmosphere (van Loon *et al.* 2011; Helwig *et al.* 2014).

Insoluble white efflorescence and other surface crusts presumed to be metal soaps have been observed on many of the W&N swatches, most notably the earliest ones from the 1940s. Examples include light brown crusts on Brown Madder (1945) and Venetian Red (1947 and 1949), and white surface crusts on Cobalt Violet (1945) and Emerald Green (1949). The crusts altered the colour of the samples significantly, were sometimes unevenly distributed, and were associated with apparently medium-rich regions at the paint surface. Swatches of Viridian (1945, 1947, 1957, 1963, 1967, 1970, 1973, 1988, 1993, 2003) made with PG18 (hydrated chromium(III) oxide) showed degradation with distinct visual characteristics not seen with other colours (Figure 5). These samples exhibited an unusual surface texture characterised by uniform, circular protrusions that were seemingly medium-rich and associated with a light green UV fluorescence. This surface texture, only seen in the thicker portion of the paint swatches, was most pronounced in the earliest samples, although the degradation also appeared to be developing in samples dating to 2003. As is the case for historic oil paintings, the growth or migration of aggregates of metal soaps may cause protrusions at paint surfaces, producing a granulated and disrupted surface. It is therefore possible that this is a manifestation of metal soap migration and aggregation, combined with pigment-medium separation, which requires further investigation.

The formation of zinc soap aggregates, protrusions, increasing transparency of paint

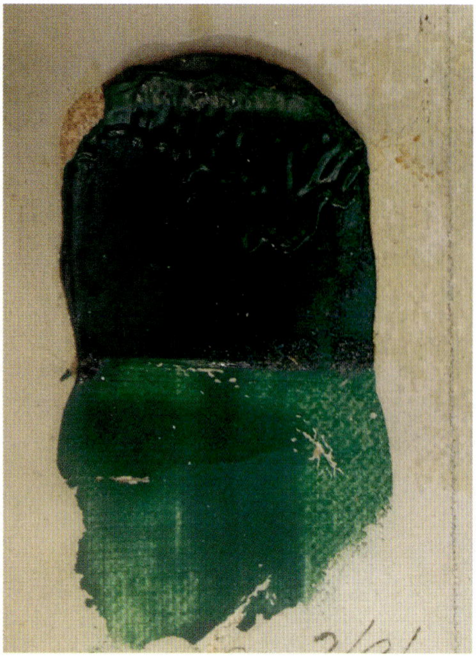

Figure 6 Medium separation on a Permanent Green W&N oil swatch from 1948 (© Tate 2015).

films, as well as structural concerns were discussed in a review on the use of zinc white in oil paint (Osmond 2012). Calcium soap protrusions have also been identified in a painting by Amiet (1868–1961) (Ferreira *et al.* 2011). Metal soaps may also cause changes in the transparency of the medium, with both decreasing and increasing transparency potentially causing darkening (Shimadzu *et al.* 2008; van den Berg *et al.* 2002). Increasing transparency in Old Master paintings containing lead white has been documented (Higgitt *et al.* 2003; van Loon *et al.* 2012). Further research is under way through the 'Paint Alterations in Time' (PAinT) project,[13] including investigations relating the internal condition of paint layers to pigment type.

Medium skins and medium drips

Pigment–medium separation is generally associated with pigments that are hygroscopic and poorly wetting (Mills *et al.* 2008). Pigment–medium separation associated with ultramarine blue swatches was noted in samples dating from 1949 to 1993, and was once again most pronounced in the early samples. Ultramarine is also prone to greyish-white discoloration, although this was not seen in the W&N samples. Recent investigations have suggested one mechanism involving disruption of six-coordinate aluminium compounds, and subsequent alteration of the chromophore, resulting in colour loss (Cato *et al.* 2016).

Related to pigment–medium separation is the observation of medium skins, which have been associated with water-sensitive 20th-century oil paints (Burnstock *et al.* 2006). A combination of cross-section examination and surface imaging has identified medium skins on various water-sensitive W&N swatches, located over thicker paint sections, often yellowed, and associated with a yellow-blue UV fluorescence. The development of oily exudates on contemporary paintings by Oldenhof has been related to the drying properties of safflower oil (Bayliss *et al.* 2016). While W&N paint formulations dating after 1966 often indicate the use of safflower oil, many of the medium skins observed in the earlier W&N samples occur in paints where poppy and linseed oil would have been used. Investigating the formation of medium skins, water sensitive or otherwise, for several different oils and paints would be useful in understanding these important surface changes.

Some of the Cadmium Green (1948) (PY35 and PG18), Permanent Green (1948) (Figure 6) (PY138, PG7, PW6), Permanent Green Light (1949) (PY74, PW6, PG7) and Yellow Ochre (1949) swatches also featured drips of discoloured medium. This could have been initiated through the pigment–medium separation process and/or the use of a high proportion of oil, which is thought to be a causal factor in the development of sticky, oily exudates noted in paintings by Jean-Paul Riopelle and Paul-Émile Borduas dating from the 1950s–70s

(Corbeil *et al.* 2004). Dripping or liquefying paints have also been observed in oil paintings of the 20th and 21st centuries (Boon and Hoogland 2014). The use of semi-drying oils is thought to be a major factor in dripping paints in many cases (Boon and Hoogland 2014; Schultz 2011), where the dripping portion appears to be enriched with polar oxidative degradation products, suggestive of polarity-driven phase separation (Hoogland *et al.* 2006). The development of treatment options for dripping paints is ongoing. So far attempts have involved the injection of stabilising metal ions into dripping paint (Boon and Hoogland 2014); curing wet paint using gamma rays (Schultz 2011); heating liquefied oil paints to 70–80°C (Franken *et al.* 2014); and painting on solvent-borne acetate driers (Bronken *et al.* 2015).

Conclusions

Colour change in modern paints is a relatively new area of research that is inextricably linked to understanding the degradation of works of art and informing conservation treatment. Colour change can be caused by many of the same phenomena explored in historic paints such as: pigment fading/discoloration due to light-induced and/or other types of chemical change; the formation of surface layers caused by physical and chemical degradation of paint materials; the migration and degradation of inherent materials to paint surfaces; dirt deposition; and the yellowing of organic materials such as oils, additives and synthetic resins. The key difference however is that modern and contemporary paintings are largely unvarnished and the qualities of paint(ing) surfaces can be particularly significant, thereby necessitating careful, informed consideration of any change or surface phenomena noted.

Largely due to quality control measures adopted by paint manufacturers, colour change in acrylic dispersion paints tends to be minimal, and primarily involves alterations to paint surfaces rather than the wholesale fading of pigments and yellowing, although these issues do occur, particularly in relation to early paint formulations, the use of gel media and artist-applied coatings. Acrylic paints can also be affected by the presence of high levels of additives, migration of inherent materials to paint surfaces in addition to passive soiling, all which can have a lasting effect on colour. With respect to formulation, early acrylic paints may also include pigments with questionable lightfastness, and additive proportions may not have been finely tuned, hence research in this area should continue to elucidate the subtleties of these changes, as well as explore specific instances of colour change.

In addition to established forms of colour change such as yellowing, modern oil paints are posing serious challenges to conservators and scientists alike with several interrelated phenomena leading to a range of changes that can also significantly affect colour as well as enhance vulnerability to mechanical and solvent action and also compromise physical properties. The formation of soft medium skins, surface crusts and metal soap formation can be particularly problematic. Modern oil paint research should draw from historic oil paint research where relevant[14] and consider the significance of the painting surface, while also increasing efforts towards understanding the range of complex phenomena that are specific to unvarnished 20th- and 21st-century oil paintings.

Acknowledgements

For co-supervision of Judith Lee, the authors would like to thank Professor Aviva Burnstock, Courtauld Institute of Art; and Ian Garrett (former W&N chemist), Patricia Smithen (collaborative doctoral student, Tate) and Spike Bucklow, Hamilton Kerr Institute for access and information pertaining to Winsor & Newton paints.

Notes

1. Except for fluorescent colours, which are not considered lightfast (Jones *et al.* 2005: 10).
2. Polyvinyl acetate and alkyd paints have not been included in this paper due to space restrictions.
3. For an overview of the development of lightfastness testing methods and regulations see Pugh and Guthrie 2001.
4. See superseded BS 1006:1990, replaced by a group of BS ISO EN standards. Available at: http://shop.bsigroup.com/ProductDetail/?pid=000000000001399764 (accessed 15 March 2016).
5. See the paper by Townsend, in this vol.
6. Some have additional listed information such as: (i) 'A' rated in full strength may fade in thin washes; (ii) Cannot be relied upon to withstand damp; (iii) Bleached by acids, acidic atmospheres; (iv) Fluctuating colour: fades in light, recovers in dark; (v) Should not be prepared in pale tints with Flake White, as these will fade; and (vi) 'A' rated with a coating of fixative.
7. www.colour-index.com/ (accessed 15 March 2016).
8. See the paper by Stenger *et al.*, in this vol.
9. Ian Garrett, Tate, London, pers. comm., 31 July 2015.
10. See www.martinsenour.com/paint-with-confidence/issues/surfactant-leaching/ (accessed 20 January 2016).
11. See www.tate.org.uk/about/projects/cleaning-modern-oil-paints (accessed 20 January 2016). J. Chung, (2016), unpublished Courtauld student project Courtauld Institute, London and Tate, London. Papers in *Issues in Contemporary Oil Paint*, edited by van den Berg *et al.* (2014): Sawicka *et al.* (311–32), Franken *et al.* (333–49), Daudin-Schotte and van Keulen (363–72), Mendes *et al.* (373–88) and Volk and van den Berg (389–406).
12. Magnesium stearates are used by W&N. Ian Garrett, pers. comm., 2015.
13. www.s4a-paint.uva.nl/ (accessed 15 March 2016).
14. See the paper by Spring, in this vol.

References

Åckerlund, L. (2012) *Efflorescence: An Investigation of Selected Paintings from the 19th to the 21st Century with a Preliminary Experimental Study of the Role of Moisture in the Development of Efflorescence*, unpublished dissertation, Courtauld Institute of Art, University of London.

ASTM International (1998) *ASTM Standards: A Century of Progress*, ASTM 1898–1998. Available at: www.astm.org/IMAGES03/Century_of_Progress.pdf (accessed 15 March 2016).

ASTM (2010) D4303-10, *Standard Test Methods for Lightfastness of Colorants used in Artists' Materials*, ASTM International, West Conshohocken, PA, 2010. Available at: www.astm.org (accessed 15 March 2016).

Aufdermarsh, C. (1988) 'The analysis of two triptychs in the Rothko Chapel', in *American Institute for Conservation Paintings Speciality Group Postprints*. Washington DC: American Institute for Conservation, 10–12.

Barker, R. and Ormsby, B. (2015) 'Conserving Mark Rothko's *Black on Maroon* 1958: the construction of a "representative sample" and the removal of graffiti ink', *Tate Papers* 23. Available at: www.tate.org.uk/research/publications/tate-papers/23/conserving-mark-rothkos-black-on-maroon-1958-the-construction-of-a-representative-sample-and-the-removal-of-graffiti-ink (accessed 15 March 2016).

Bayliss, S., van den Berg, K.J., Burnstock, A., de Groot, S., van Keulen, H. and Sawicka, A. (2016) 'An investigation into the separation and migration of oil in paintings by Erik Oldenhof', *Microchemical Journal* 124: 974–82.

Berrie, B. (ed.) (2007) *Artists' Pigment: A Handbook of their History and Characteristics*, vol. 4. London: Archetype Publications.

Berrie, B.H. and Quillen-Lomax, S (1997) 'Azo pigments: their history, synthesis, properties and use in artists' materials', in *Conservation Research 1996/1997*, Studies in the History of Art, vol. 57. Washington DC: National Gallery of Art, 9–33.

Blumenroth, D., Zumbuhl, S., Scherrer, N.C. and Müller, W. (2014) 'Sensitivity of modern oil paints to solvents: effects on synthetic organic pigments', in K.J. van den Berg, M. de Keijzer, A. de

Tagle, J. Kreuger, G. Heydenreich, A. Burnstock and T. Learner (eds), *Issues in Contemporary Oil Paint*. Cham: Springer International Publishing, 351–62.

Boon, J. J. and Hoogland, F.G. (2014) 'Investigating fluidizing dripping pink commercial paint on van Hemert's *Seven-Series Works* from 1990–1995', in K.J. van den Berg, M. de Keijzer, A. de Tagle, J. Kreuger, G. Heydenreich, A. Burnstock and T. Learner (eds), *Issues in Contemporary Oil Paint*. Cham: Springer International Publishing, 227–46.

Boon, J. J., Peulvé, S., van den Brink, O., Duursma, M. and Rainford, D. (1997) 'Molecular aspects of mobile and stationary phases in ageing tempera and oil paint films', in T. Bakkenist, R. Hoppenbrouwers and H. Dubois (eds), *Early Italian Paintings: Techniques and Analysis*. Maastricht: Limburg Conservation Institute (SRAL), 35–56.

Bronken, I., Boon, J. and and Vila, A. (2015) 'In search of a solution for softening and weeping paints in contemporary paintings by Tal R and Frank van Hemert', in A. Barros d'Sa, L. Bone, R. Clarricoates and H. Dowding (eds), *Current Technical Challenges in the Conservation of Paintings*. London: Archetype Publications, 28–41.

Burnstock, A. and van den Berg, K.J. (2014) 'Twentieth century oil paint: the interface between science and conservation and the challenges for modern oil paint research', in K.J. van den Berg, M. de Keijzer, A. de Tagle, J. Kreuger, G. Heydenreich, A. Burnstock and T. Learner (eds), *Issues in Contemporary Oil Paint*. Cham: Springer International Publishing, 1–19.

Burnstock, A., Caldwell, M. and Odlyha, M. (1993) 'A technical examination of surface deterioration of Stanley Spencer's paintings at Sandham Memorial Chapel', in J. Bridgland (ed.), *Preprints of the ICOM-CC 10th Triennial Meeting*, vol. 1. London: James and James, 231–8

Burnstock, A., van den Berg, K.J., de Groot, S. and Wijnberg, L. (2006) 'An investigation into water sensitive oil paints in twentieth-century paintings', in T. Learner, J.W. Smithen, J. Krueger and M.R. Schilling (eds), *Modern Paints Uncovered*. Los Angeles: Getty Conservation Institute, 177–88.

Burnstock, A., Lee, J., Ormsby, B. and van den Berg, K.J. (2015) 'Water sensitivity of modern oil paint films', in *CESMAR7 VII Congresso Internazionale Coloure e Conservazione Politecnico di Milano 13–14 November 2015* (in press).

Caldwell, M. (2002) *Some Developments in British Paint Manufacture c.1850–2000 with Reference to Surface Deterioration on Paintings and in Particular Free Fatty Acid Efflorescence*, unpublished report for the National Trust. London: Courtauld Institute of Art.

Carlyle, L. (2001) *The Artist's Assistant: Oil Painting Instruction Manuals and Handbooks in Britain 1800–1900 with Reference to Selected Eighteenth-century Sources*. London: Archetype Publications.

Carlyle, L., Binnie, N. and Kaminska, E. (2002) 'The yellowing/bleaching of oil paintings and oil paint samples, including the effect of oil processing, driers, and mediums on the colour of lead white paint', in R. Vontobel (ed.), *Preprints of the ICOM-CC 13th Triennial Meeting*, vol. 1. London: James and James, 328–37.

Carlyle, L., Boon, J., Bustin, M. and Smithen, P. (2008) 'The substance of things', in A. Borchardt-Hume (ed.), *Rothko*. London: Tate Publishing, 75–87.

Cato, E., Borca, C., Huthwelker, T. and Ferreira, E. (2016) 'Aluminium X-ray absorption near-edge spectroscopy analysis of discoloured ultramarine blue in 20th century oil paintings', *Microchemical Journal* 126: 18–24.

Christie, R.M. (2002) *The Organic and Inorganic Chemistry of Pigments*. London: Oil and Colour Chemists Association.

Chroma Australia (2016) *Lightfastness Rating – What it Means*. Available at: http://www.chromaonline.com/products/au/archival_oils/colour_range/lightfastness_rating_what_it_means (accessed 13 January 2016).

Corbeil, M.C., Helwig, K. and Poulin, J. (2004) 'Analysis of the painted oeuvre of Jean-Paul Riopelle: from oil to mixed media', in A. Roy and P. Smith (eds), *Modern Art, New Museums*. London: International Institute for Conservation of Historic and Artistic Works, 170–73.

Defeyt, C. and Strivay, D. (2014) 'PB15 as 20th and 21st century artists' pigments: conservation concerns', *e-Preservation Science* 11: 6–14.

Doménech-Carbó, M.T., Silva, M.F., Aura-Castro, E. *et al.* (2011) 'Study of behaviour on simulated daylight ageing of artists' acrylic and poly(vinyl acetate) paint films', *Analytical and Bioanalytical Chemistry* 399: 3155–304.

Eastaugh, N. (1990) 'The visual effects of dirt on paintings', in S. Hackney, J. Townsend and N. Eastaugh (eds), *Dirt and Pictures Separated.* London: UKIC, 51–5.

Ferreira, E., Boon, J.J., Stampanoni, M. and Marone, F. (2011) 'Study of the mechanism of formation of calcium soaps in an early 20th-century easel painting with correlative 2D and 3D microscopy', in J. Bridgland (ed.), *Preprints of the ICOM-CC 16th Triennial Meeting.* Almada: Critério, 1604, 1–8.

Franken, V., Schultz, J., Heydenreich, G., Jägers, E., Müller, W. and Zumbühl, S. (2014) 'Set back the race: treatment strategies for running oil paint', in K.J. van den Berg, M. de Keijzer, A. de Tagle, J. Kreuger, G. Heydenreich, A. Burnstock and T. Learner (eds), *Issues in Contemporary Oil Paint.* Cham: Springer International Publishing, 333–49.

Fremout, W., Verboven, M. and Saverwyns, S. (2014) 'Accelerated tobacco smoke staining on waterborne acrylic paintings caused by exuding surfactants: a study with Py-GC/MS and THM-GC/MS', *e-Preservation Science* 11: 47–53.

Gavett, B. (2012) 'Artists' paints', in J.V. Koleske (ed.), *Paint and Coating Testing Manual*, 15th edn. West Conshohocken PA: ASTM International, 768.

Gherlardi, E., Degano, I., Colombini, M.P., Mazurek, J., Schilling, M., Khanjian, H. and Learner, T. (2015) 'A multi-analytical study on the photochemical degradation of synthetic organic pigments', *Dyes and Pigments* 123: 396–403.

Golden Artist Colors (1995) 'How to test for lightfastness', 1 January 1995. Available at: www.justpaint.org/how-to-test-for-lightfastness (accessed 21January 2016).

Golden, M. (2010a) 'Following the research on the acrylic medium', *Just Paint*, 1 September 2010. Available at: www.justpaint.org/following-the-research-on-the-acrylic-medium/ (accessed 19 January 2016).

Golden, M. (2010b) 'Defining the acrylic patina', *Just Paint*, 1 September 2010. Available at: www.justpaint.org/defining-the-acrylic-patina/ (accessed 19 January 2016).

Herbst, W. and Hunger, K. (2004) *Industrial Organic Pigments.* Weinheim: Wiley-VCH.

Helwig, K., Poulin, J., Corbeil, M-C., Moffat, E. and Duguay, D. (2014) 'Conservation issues in several twentieth-century Canadian oil paintings: the role of zinc carboxylate reaction products', in K.J. van den Berg, M. de Keijzer, A. de Tagle, J. Kreuger, G. Heydenreich, A. Burnstock and T. Learner (eds), *Issues in Contemporary Oil Paint.* Cham: Springer International Publishing, 167–84.

Higgitt, C., Spring, M. and Saunders, D. (2003) 'Pigment-medium interactions in oil paint films containing red lead or lead-tin yellow', *National Gallery Technical Bulletin* 24: 75–95.

Hillary, S. and Ormsby, B. (2014) 'Surface streaking on a matt PVA-acrylic painting', poster 1018-500 presented at the *ICOM-CC 17th Triennial Conference, Melbourne, 2014.*

Hinde, L., van den Berg, K. J., de Groot, S. and Burnstock, A. (2011) 'Characterisation of surface whitening in 20th century European paintings at Dudmaston Hall, United Kingdon', in J. Bridgland (ed.), *Preprints of the ICOM-CC 16th Triennial Meeting.* Almada: Critério, 1310, 1–10.

Hoogland, F., van der Horst, J. and Boon, J.J. (2006) 'Liquefying oil paint in some late-twentieth-century paintings', in T. Learner, J.W. Smithen. J. Krueger and M.R. Schilling (eds), *Modern Paints Uncovered.* Los Angeles: Getty Conservation Institute, 282–3.

Izzo, F.C., Balliana, E., Pinton, F. and Zendri, E. (2014) 'Preliminary study of the composition of commercial oil, acrylic and vinyl paints and their behaviour after accelerated ageing conditions', *Conservation Science in Cultural Heritage* 14: 353–69.

Jablonksi, E., Learner, T., Hayes, J. and Golden, M. (2004) 'Conservation concerns for acrylic emulsion paints: a literature review', *Tate Papers* 2. Available from: www.tate.org.uk/download/file/fid/7414 (accessed 19 March 2016).

Jones, F.N., Maoa, W., Ziemer, P.D., Xiao, F., Hayes, J. and Golden, M. (2005) 'Artist paints: an overview and preliminary studies of durability', *Progress in Organic Coatings* 52(1): 9–20.

Keijzer, M. de (2014) 'The delight of modern organic pigment creations', in K.J. van den Berg, M. de Keijzer, A. de Tagle, J. Kreuger, G. Heydenreich, A. Burnstock and T. Learner (eds), *Issues in Contemporary Oil Paint*. Cham: Springer International Publishing, 75–104.

Kennedy, A.R., Stewart, H., Eremin, K. and Stenger, J. (2012) 'Lithol red: a systematic structural study on salts of a sulphonated azo pigment', *Chemistry: A European Journal* 18: 3064–9.

Keune, K., Hoogland, F., Boon, J. J., Peggie, D. and Higgitt, C. (2008) 'Comparative study of the effect of traditional pigments on artificially aged oil paint systems using complementary techniques', in J. Bridgland (ed.) *Preprints of the ICOM-CC 15th Triennial Meeting*, vol. 2. New Delhi: Allied Publishers Pvt Ltd, 22–6.

Koller, J. and Burmester, A. (1990) 'Blanching of unvarnished modern paintings: a case study on a painting by Serge Poliakoff', in J.S. Mills and P. Smith (eds), *Cleaning, Retouching and Coatings*. London: IIC, 138–43.

Learner, T. (2005) *The Analysis of Modern Paints*. Los Angeles: Getty Publications.

Learner, T., Chiantore, O. and Scalarone, D. (2002) 'Ageing studies of acrylic emulsion paint', in R. Vontobel (ed.), *Preprints of the ICOM-CC 13th Triennial Meeting*, vol. 2. London: James and James, 911–19.

Levison, H.W. (1976). *Artists' Pigments: Lightfastness Tests and Ratings*. Florida: Colorlab.

Levison, H.W. (1985) 'Yellowing and bleaching of paint films', *Journal of the American Institute for Conservation* 24(2): 69–76.

Mallégol, J., Lemaire, J. and Gardette, J.-L. (2001) 'Yellowing of oil-based paints', *Studies in Conservation* 46(2): 121–31.

Mass, J., Sedlmair, J., Patterson, C.S., Carson, D., Buckley, B. and Hirschmugl, C. (2013) 'SR-FTIR imaging of the altered cadmium sulfide yellow paints in Henri Matisse's *Le Bonheur de vivre* 1905–6: examination of visually distinct degradation regions', *The Analyst* 138: 20.

Michalski, S. (1991) 'Paintings: their response to temperature, relative humidity, shock and vibration', in M.F. Mecklenburg (ed.), *Works of Art in Transit*. Washington DC: National Gallery of Art, 223–48.

Mills, L., Burnstock, A., Duarte, F., de Groot, S., Megens, L., Bisschoff, M. and van den Berg, K.J. (2008) 'Water sensitivity of modern artists' oil paints', in J. Bridgland (ed.) *Preprints of the ICOM-CC 15th Triennial Meeting*. New Delhi: Allied Publishers Pvt Ltd, 651–9.

Monico, L., Janssens, K., Cotte, M., Sorace, L., Vanmeert, F., Brunetti, B. and Miliani. C. (2016) 'Chromium speciation methods and infrared spectroscopy for studying the chemical reactivity of lead chromate-based pigments in oil medium', *Microchemical Journal* 124: 272–82.

Myers, D.G. (2013) *Color of Art Pigment Database*. Available at: www.artiscreation.com/Color_index_names.html (accessed 14 January 2016).

Ormsby, B., Learner, T., Foster, G., Druzik, J. and Schilling, M. (2007) 'Wet-cleaning acrylic emulsion paint films: an evaluation of physical, chemical and optical changes', in T. Learner, J.W. Smithen, J. Krueger and M.R. Schilling (eds), *Modern Paints Uncovered*. Los Angeles: Getty Conservation Institute, 187–98.

Ormsby, B., Smithen, P., Hoogland, F., Learner, T. and Miliani, C. (2008) 'A scientific investigation into the surface cleaning of acrylic emulsion paintings', in J. Bridgland (ed.) *Preprints of the ICOM-CC 15th Triennial Meeting*, vol. 2. New Delhi: Allied Publishers Pvt Ltd, 857–65.

Ormsby, B., Cross, M., Kampasakali, E., Chantrier Aasen, L. and Smithen, P. (2011) 'A preliminary evaluation of artists' and conservation varnishes for acrylic emulsion paint films', in J. Bridgland (ed.), *Preprints of the ICOM-CC 16th Triennial Meeting*. Almada: Critério, 1318, 1–11.

Ormsby, B., Hodgkins, R. and von Aderkas, N. (2012) 'Preliminary investigations into two new acrylic emulsion paint formulations: W&N Artists' Acrylic Colours and Golden Open Acrylics', *e-Preservation Science* 9: 9–16.

Ormsby, B., Kampasakali, E. and Learner, T. (2013) 'Surfactants and acrylic emulsion paints: evaluating changes induced by wet surface cleaning treatments', in M.F. Mecklenburg, A.E. Charolda and R.J. Koestler (eds), *New Insights into the Cleaning of Paintings*. Washington DC: Smithsonian Institution Scholarly Press, 159–64.

Ormsby, B., Learner, T., Crook, J. and Smithen, P. (2014) 'John Hoyland: a British master of acrylic painting', in A. Sgamellotti, B.G. Brunetti and C. Miliani (eds), *Science and Art: The Painted Surface*. London: RSC Publishing, 522–45.

Osmond, G. (2012) 'Zinc white: a review of zinc oxide pigment properties and implications in oil-based paintings', *AICCM Bulletin* 33(1): 20–29.

Osmond, G. (2014) 'Zinc white and the influence of paint composition for stability in oil based media', in K.J. van den Berg, M. de Keijzer, A. de Tagle, J. Kreuger, G. Heydenreich, A. Burnstock and T. Learner (eds), *Issues in Contemporary Oil Paint*. Cham: Springer International Publishing, 263–82.

Perry, R. (1990) 'Problems of dirt accumulation and its removal from unvarnished paintings: a practical review', in S. Hackney, J. Townsend and N. Eastaugh (eds), *Dirt and Pictures Separated*. London: UKIC, 3–6.

Pintus, V. and Schreiner, M. (2011) 'Characterization and identification of acrylic binding media: influence on UV light on the ageing process', *Anaytical and Bioanalytical Chemistry* 399: 2961–76.

Pintus, V., Wei, S. and Schreiner, M. (2015) 'Accelerated UV ageing studies of acrylic, alkyd, and polyvinyl acetate paints: influence of inorganic pigments', *Microchemical Journal* 121: 949–69.

Pouyet, E., Cotte, M., Favard, B., Salome, M., Meirer, F., Mehta, A., Uffelman, E.S., Hull, A., Vanmeert, F., Kieffer, J., Burghammer, M., Janssens, K., Sette, F. and Masset, J. (2015) '2D-X-ray and FTIR micro-analysis of the degradation of cadmium yellow pigments in paintings of Henri Matisse', *Applied Physics A* 121(2): 967–80.

Pugh, S.L. and Guthrie, J.T. (2001) 'The development of light fastness testing and light fastness standards', *Review of Progress in Coloration* 31: 42–56.

Quillen-Lomax, S. (2010a) 'The application of X-ray powder diffraction for the analysis of synthetic organic pigments. Part 1: Dry pigments', *Journal of Coatings Technology and Research* 7(3): 331–46.

Quillen-Lomax, S. (2010b) 'The application of X-ray powder diffraction to the analysis of synthetic organic pigments. Part 2: Artists' paints', *Journal of Coatings Technology and Research* 7(3): 325–30.

Quillen-Lomax, S. and Learner, T. (2006) 'A review of the classes, structures, and methods of analysis of synthetic organic pigments', *Journal of the American Institute for Conservation* 45: 107–26.

Quillen-Lomax, S., Schilling, M.R and Learner, T. (2007) 'The identification of synthetic organic pigments by FTIR and DTMS', in M. Greenberg (ed.), *Modern Paints Uncovered*. Los Angeles: Getty Conservation Institute, 105–17.

Quillen-Lomax, S., Lomax, J.F. and DeLuca-Westrate, A. (2014) 'The use of Raman microscopy and laser desorption ionization mass spectrometry in the examination of synthetic organic pigments in modern works of art', *Journal of Raman Spectroscopy* 45(6): 448–55.

Rakoff, H., Thomas, F.L. and Gast, L.E. (1979) 'Reversibility of yellowing phenomenon in linseed based paints', *Journal of Coatings Technology* 51(649): 25–8.

Rie, R. de la (1982) 'Fluorescence of paint and varnish layers (Part III)', *Studies in Conservation* 27(3): 102–8.

Sands, S. (2007) 'The nomenclature of color', *Just Paint*, 1 January 2007. Available at: www.justpaint.org/the-nomenclature-of-color/ (accessed 13 January 2016).

Sands, S. (2015) 'QoR® watercolor questions: labeling and lightfastness ratings', *Just Paint*, 1 February 2015. Available at: www.justpaint.org/qor-watercolor-questions-labeling-and-lightfastness-ratings/ (accessed 8 January 2016).

Saunders, D. and Kirby, J. (2004) 'The effect of relative humidity on artists' pigments', *National Gallery Technical Bulletin* 25: 62–72.

Sawicka, A., Burnstock, A., Izzo, F.C., Keune, K., Boon, J.J., Kirsch, K. and van den Berg, K.J. (2014) 'An investigation into the viability of removal of lead soap efflorescence from contemporary oil paintings', in K.J. van den Berg, M. de Keijzer, A. de Tagle, J. Kreuger, G. Heydenreich, A. Burnstock and T. Learner (eds), *Issues in Contemporary Oil Paint*. Cham: Springer International Publishing, 311–32.

Scherrer N.C., Stefan, Z., Francoise, D., Annette, F. and Renate, K. (2009) 'Synthetic organic pigments of the 20th and 21st century relevant to artist's paints: Raman spectra reference collection', *Spectrochimica Acta Part A* 73: 505–24.

Schminke (2013) 'Lightfastness – a measure of the durability of artists' colours'. Available at: www.schmincke.de/fileadmin/Produktinformationen_Englisch_2014/Good_to_know/PI_lightfastness_130606.pdf. (accessed 8 January 2016).

Schultz, J. (2011) 'Liquefying oil paint: contemporary paintings in danger', in J. Bridgland (ed.), *Preprints of the ICOM-CC 16th Triennial Meeting*. Almada: Critério, 1320.

Shepherd, C. (2014) *An Investigation into the Materials, Techniques and Conservation Issues of Oil Paintings c.1942–57 by Sir Matthew Smith (1879–1959) in the Collection of the Guildhall Art Gallery, City of London*, unpublished dissertation, Courtauld Institute of Art, University of London.

Shimadzu, Y., Keune, K. and van den Berg, K.J. (2008) 'The effects of lead and zinc white saponification on surface appearance of paint', in J. Bridgland (ed.), *Preprints of the ICOM-CC 16th Triennial Meeting*, vol. 2. Almada: Critério, 626–32.

Singer, B., Devenport, J. and Wise, D. (1995) 'Examination of blooming problem in a collection of unvarnished paintings', *The Conservator* 19: 3–9.

Smith, G.D. (2007) 'Aging characteristics of a contemporary acrylic emulsion used in artists' paints', in T. Learner, J.W. Smithen, J. Krueger and M.R. Schilling (eds), *Modern Paints Uncovered*. Los Angeles: Getty Conservation Institute, 236–46.

Standeven, H.A.L. (2008) 'The history and manufacture of lithol red, a pigment used by Mark Rothko in his Seagram and Harvard murals of the 1950s and 1960s', *Tate Papers* 10. Available at: www.tate.org.uk/download/file/fid/7323 (accessed 19 March 2016).

Stenger, J., Kwan, E., Eremin, K., Speakman, S., Kirby, D., Stewart, H., Huang, S., Kennedy, A., Newman, R. and Khandekar, N. (2010) 'Lithol red salts: characterization and deterioration', *e-Preservation Science* 7(2): 147–57.

Townsend, J.H., Ormsby. B., Barker, R. and Smithen, P. (2014) 'A microfading study of Mark Rothko's materials for the Seagram murals (1958–59)', poster 1011-356 presented at the *ICOM-CC 17th Triennial Conference, Melbourne, 2014*.

Tumosa, C.S. (2001) 'A brief history of aluminium stearate as a component of paint', *WAAC Newsletter* 23(3).

van den Berg, K.J., Boon, J.J. and van den Berg, J.J. (1999) 'Chemical changes in curing and ageing oil paints', in D. Grattan (ed.), *Preprints of the ICOM-CC 12th Triennial Meeting*, vol. 1. Paris: ICOM, 248–53.

van den Berg, K.J., Geldof, M., de Groot, S. and van Keulen, H. (2002) 'Darkening and surface degradation in 19th- and early 20th-century paintings: an analytical study', in R. Vontobel (ed.), *Preprints of the ICOM-CC 13th Triennial Meeting*, vol. 1. London: James and James, 464–72.

van den Berg, K.J., de Keijzer, M., de Tagle, A., Kreuger, J., Heydenreich, G., Burnstock, A. and Learner, T. (eds) (2014) *Issues in Contemporary Oil Paint*. Cham: Springer International Publishing.

van Loon, A., Noble, P. and Boon, J.J. (2011) 'White hazes and surface crusts in Rembrandt's *Homer* and related paintings', in J. Bridgland (ed.), *Preprints of the ICOM-CC 16th Triennial Meeting*. Almada: Critério, 1–10.

van Loon, A., Noble, P. and Burnstock, A. (2012) 'Ageing and deterioration of traditional oil and tempera paints', in J.H. Stoner and R. Rushfield (eds), *The Conservation of Easel Paintings*, Abingdon: Routledge, 214–41.

Whitmore P.M. and Cass, G.R. (1989) 'The fading of artists' colorants by exposure to atmospheric nitrogen dioxide', *Studies in Conservation* 34(2): 85–97.

Winsor & Newton (2016a) *Composition and Permanence Tables*. Available at: www.winsornewton.com/uk/discover/resources/composition-permanence (accessed 8 January 2016).

Winsor & Newton (2016b). *Professional Acrylic*. Available at: www.winsornewton.com/uk/discover/resources/colour-charts/professional-acrylic (accessed 19 January 2016).

Zhang, A. (2013) *Photocatalytical Effect of TiO2 Pigments on the Surface of Paint Films*, MSc thesis, Delft University of Technology.

Available at: http://repository.tudelft.nl/view/ir/uuid%3A940a74cf-2f11-45af-a4d5-d66218632535/ (accessed 30 March 2016).

Authors' addresses

- Bronwyn Ormsby, Principal Conservation Scientist, Tate Britain, London, UK (bronwyn.ormsby@tate.org.uk)
- Judith Lee, AHRC Collaborative Doctoral Student, Courtauld Institute of Art and Tate, London, UK (judith.lee@tate.org.uk)

Non-invasive colour restoration of Mark Rothko's Harvard murals using light from a digital projector

Jens Stenger, Narayan Khandekar, Ramesh Raskar, Santiago Cuellar, Ankit Mohan, Rudolf Gschwind, Mary Schneider Enriquez, Carol Mancusi-Ungaro and Christina Rosenberger

A series of five paintings, known as Mark Rothko's Harvard murals, was painted in 1962 and installed in the Holyoke Center, Harvard University, in January 1964 (Cohn 1988). The paintings changed colour due to the presence of lithol red (Hensick and Whitmore 1988; Stenger *et al.* 2010, Kennedy *et al.* 2012), a fugitive red pigment, and excessive exposure to natural light in a room with large windows for more than 15 years. They were removed from display to dark storage in 1979 and rarely displayed. Reinstating the faded colour using traditional painting restoration methods would almost certainly have resulted in a non-reversible treatment due to the matt and unvarnished nature of paint surfaces, therefore the treatment possibilities were considerably limited.

An innovative treatment using light from a digital projector to compensate for the lost colour on a pixel-by-pixel basis was recently developed (Cuellar *et al.* 2011; Stenger *et al.* 2015a) in an extension of Lafontaine's light experiments of 1986 (Lafontaine 1986). The original colour of the works was determined by the digital restoration of Kodak Ektachrome photographs taken in 1964 (Frey and Gschwind 1994; Gschwind and Frey 1994), and also by comparison to unfaded passages of all canvases, including a sixth mural which was never installed at Harvard. With a camera-projector system (Grossberg *et al.* 2004), a compensation image was calculated – in effect a map of the lost colour over more than 2 million pixels. The compensation image was then aligned and projected onto the original canvas resulting in a restored colour appearance. It was found that blending additional ambient light with the colour-correcting light projection gave the best results. This was repeated for each of the five paintings. As a result the original colour appearance was created without physically altering the painting and is therefore completely reversible. In fact, the unrestored and colour-corrected object can easily be compared by switching the projector on or off. However, with the projectors off the paintings are under low illumination compared to the rest of the room and the human eye perceives them, in this state, as very dark. With the digital projector's colour-correcting illumination the group of murals now works together in a way that has not been possible for many decades. Overall lighting and architecture played a key role in the treatment of the mural cycle as an environment. In addition, a detailed examination

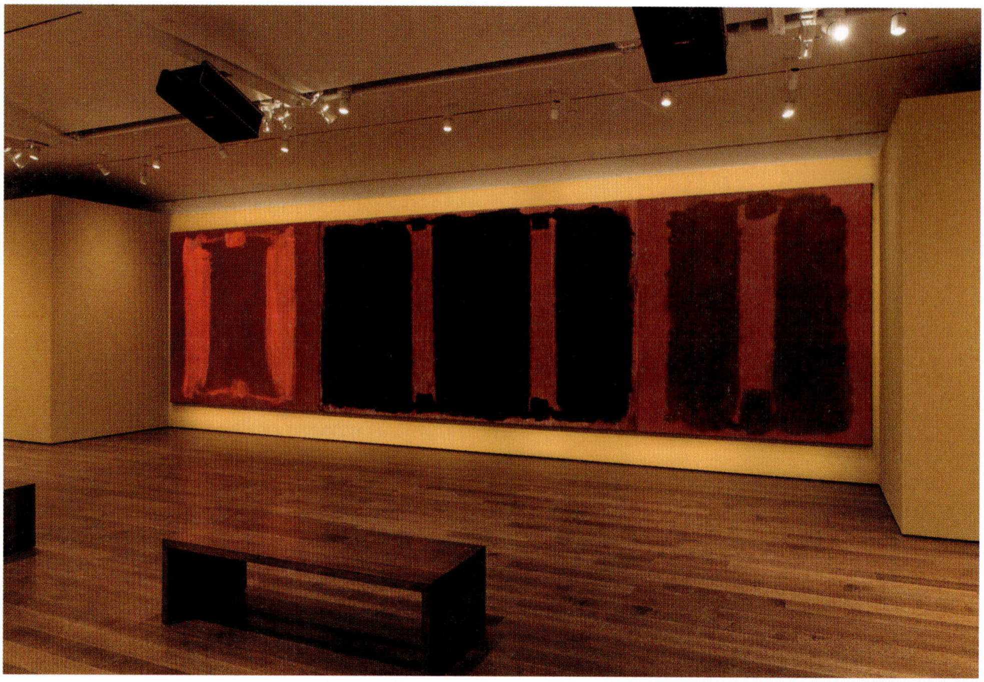

Figure 1 Colour restored Panel One, Panel Two, and Panel Three (Harvard Mural Triptych) by Mark Rothko, 1962. (Photo: Peter Vanderwarker; © President and Fellows of Harvard College)

of Rothko's materials and methods for these paintings was carried out to understand how the murals were made and why they have aged differently (Stenger *et al.* 2015b).

To our knowledge, this novel approach has not previously been used for the display of paintings although similar projects have been on display at Hampton Court (Perkins *et al.* 2011) and at the Brömershof in Rüdesheim (Janke 2011). Microfading tests were used to calculate the light sensitivity of the faded paintings and ensure that the paintings would not alter as a result of this treatment. Consequently, Mark Rothko's Harvard murals, which had fallen out of step with each other because of the differences in fading, can now be viewed as a single harmonious installation. The murals formed the basis of the first exhibition, *Mark Rothko's Harvard Murals*, in the temporary exhibition galleries of the newly renovated Harvard Art Museums from November 2014 until July 2015. The projectors were turned off for the last hour of each day so that the works could be seen without the colour-compensating projected light, an event that became popular with visitors.

The show and the conservation approach were discussed widely, including at a study day in March 2015 (Aberbach 2015) at the Harvard Art Museums and also in the media (Sheets 2014; Walsh 2014; Menand 2015; Bois *et al.* 2015). At the study day, the artist Terry Winter articulated that 'once the projectors were off, Rothko entered the room. I like the paintings both ways. The drama of turning off the projectors is like the move from comedy to tragedy, unexpectedly.' Brad Epley, chief conservator of the Menil Collection was 'distracted by a pillow of light in front of the canvas. The restored version is a digital remaster, and the unrestored is like a vinyl LP.' For him the exhibition provoked the questions of 'what is the most authentic version?' and 'at what point do we accept change?' The strong difference from colour-restored appearance to the state without projectors made it difficult not to see *only* the restoration according to Martha Buskirk from Montserrat College of Art: 'We have to be careful to look at the paintings and not the restoration.' Thinking conceptually about the murals, Jeffrey Weiss from the Guggenheim Museum felt that 'the substitution of the metaphorical light of the painting with the literal, projected light violates the terms of the work. This is a fundamental obstacle. Whole-cloth deceptive illusion is unnerving. The restoration produces a simulacrum, and we are in danger of substituting it for the original, and the original loses its authenticity.' More positive criticism came from the children of Mark Rothko. Kate Rothko Prizel said that 'The setting is not a problem, you experience the room. The space feels right. It feels like Rothko's luminosity.' Her brother Christopher Rothko added that he was 'really struck right away not so much by the colour but by the way they still felt like paintings. You still have the feel of the canvas. For me that's what makes it still feel so believable. Because my father's brushstrokes are still there.'

In the summer of 2015, *Artforum* published a panel discussion (Bois *et al.* 2015) on the exhibition with Yve-Alain Bois, Harry Cooper, Lynn Hershman Leeson, Carol Mancusi-Ungaro, Ken Okiishi, R.H. Quaytman, David Reed, Jeffrey Weiss and Michelle Kuo. Weiss repeated his criticism about replacing the metaphorical with the literal and in response Cooper pointed out that 'I think it's very rare to see the application of digital technology, in such a painstakingly material way, marry itself to a physical substrate. It's something I certainly have never seen before. … Here the surface was treated with a kind of love. That's the only way I can really describe it.'

References

Aberbach, K. (2015) 'Up for debate: Rothko murals, "restored" by digital lighting, are focus of two days of discussions'. Available at: http://news.harvard.edu/gazette/story/2015/03/up-for-debate/ (accessed 19 March 2016).

Bois, Y.-A., Cooper, H., Hershman Leeson, L., Mancusi-Ungaro, C., Okiishi, K., Quaytman, R.H., Reed, D., Weiss, J. and Kuo, M. (2015) 'Light repairs: a roundtable on the restoration of Mark Rothko's Harvard murals', *Artforum* Summer 2015. Available at: https://artforum.com/inprint/issue=201506&id=52269 (accessed 19 March 2016).

Cohn, M.B. (ed.) (1988) *Mark Rothko's Harvard Murals*. Cambridge MA: Center for Conservation and Technical Studies, Harvard University Art Museums.

Cuellar, S., Stenger, J., Gschwind, R., Mohan, A., Mukaigawa, Y., Raskar, R., Eremin, K. and Khandekar, N. (2011) 'Non-invasive color restoration of faded paintings using light from a digital projector', in J. Bridgland (ed.), *Preprints of the ICOM-CC 16th Triennial Meeting*. Almada: Critério [CD-ROM].

Frey, F. and Gschwind, R. (1994) 'Mathematical bleaching models for photographic three-color materials', *Journal of Imaging Science and Technology* 38(6): 513.

Grossberg, M.D., Peri, H., Nayar, S.K. and Belhumeur, P.N. (2004) 'Making one object look like another: controlling appearance using a projector–camera system', in *Computer Vision and Pattern Recognition (CVPR). Proceedings of the 2004 IEEE Computer Society Conference 27 June–2 July 2004*, vol. 1. New York: IEEE, 452–9.

Gschwind, R. and Frey, F. (1994) 'Electronic imaging: a tool for the reconstruction of faded color photographs', *Journal of Imaging Science and Technology* 38(6): 520.

Hensick, T. and Whitmore, P. (1988) 'Rothko's Harvard murals', in M.B. Cohn (ed.), *Mark Rothko's Harvard Murals*. Cambridge MA: Center for Conservation and Technical Studies, Harvard University Art Museums, 15–30.

Janke, M. (2011) *Rekonstruktion oder Ergänzung durch Lichtbildprojektion – eine Methode zur Präsentation teilweise oder komplett verloren gegangener Wandmalerei und Architekturpolychromie*, Diplomarbeit.

Kennedy, A.R., Stewart, H., Eremin, K. and Stenger, J. (2012) 'Lithol red: a systematic structural study on salts of a sulphonated azo pigment', *Chemistry* 18: 3064–9.

Lafontaine, R.H. (1986) 'Seeing through a yellow varnish: a compensating illumination system', *Studies in Conservation* 31(3): 97–102.

Menand, L. (2015) 'Watching them turn off the Rothkos', *The New Yorker* April 15, 2015. Available at: www.newyorker.com/culture/cultural-comment/watching-them-turn-off-the-rothkos (accessed 19 March 2016).

Perkins, R., Owens,H., Carr, C., Hallett, K., Gibb, I. and Frame, K. (2011) 'Henry VIII's tapestries revealed', in J. Bridgland (ed.), *Preprints of the ICOM-CC 16th Triennial Meeting*. Almada: Critério [CD-ROM].

Shea, A. (2014) 'Harvard's famously damaged Rothko paintings "restored" with light', *The ARTery*, 22 May2014 Available at: http://artery.wbur.org/2014/05/20/harvard-rothkos (accessed 19 March 2016).

Sheets, H.M. (2014) 'A return for Rothko's Harvard murals', *The New York Times*, October 23, 2014. Available at: www.nytimes.com/2014/10/26/arts/artsspecial/a-return-for-rothkos-harvard-murals-.html?_r=0 (accessed 19 March 2016).

Stenger, J., Kwan, E., Eremin, K., Speakman, S., Kirby, D., Stewart, H., Huang, S., Kennedy, A., Newman, R. and Khandekar, N. (2010) 'Lithol red salts: characterization and deterioration', *e-Preservation Science* 7(2): 147–57.

Stenger, J., Khandekar, N., Raskar, R., Cuellar, S., Mohan, A. and Gschwind, R. (2015a) 'Conservation of a room: a treatment proposal for Mark Rothko's Harvard murals', *Studies in Conservation*. Available at: http://dx.doi.org/10.1179/2047058415Y.0000000010 (accessed 19 March 2016).

Stenger, J., Khandekar, N., Wilker, A., Kallsen, K., Kirby, D.P. and Eremin, K. (2015b) 'The making of Mark Rothko's Harvard murals', *Studies in Conservation*. Available at: http://dx.doi.org/10.1179/2047058415Y.0000000009 (accessed 19 March 2016).

Walsh, C. (2014) 'A light touch for Rothko murals', *Harvard Gazette*, May 20, 2014. Available at: http://news.harvard.edu/gazette/story/2014/05/a-light-touch-for-rothko-murals/ (accessed 19 March 2016).

Authors addresses

- *Author for correspondence*: Jens Stenger, Institute for the Preservation of Cultural Heritage, Yale University, West Haven CT, USA/ Straus Center for Conservation and Technical Studies, Harvard Art Museums, Cambridge MA, USA (Jens.stenger@yale.edu)
- Narayan Khandekar, Straus Center for Conservation and Technical Studies, Harvard Art Museums, Cambridge MA, USA
- Ramesh Raskar, Camera Culture Group, Media Lab, Massachusetts Institute of Technology, Cambridge MA, USA
- Santiago Cuellar, Camera Culture Group, Media Lab, Massachusetts Institute of Technology, Cambridge MA, USA
- Ankit Mohan, Camera Culture Group, Media Lab, Massachusetts Institute of Technology, Cambridge MA, USA
- Rudolf Gschwind, Digital Humanities Lab, University of Basel, Basel, Switzerland
- Mary Schneider Enriquez, Division of Modern and Contemporary Art, Harvard Art Museums, Cambridge MA, USA
- Carol Mancusi-Ungaro, Center for the Technical Study of Modern Art, Harvard Art Museums, Cambridge MA, USA/ Conservation and Research Department, Whitney Museum of American Art, New York NY, USA
- Christina Rosenberger, Center for the Technical Study of Modern Art, Harvard Art Museums, Cambridge MA, USA

Microfading: a useful tool for paintings conservators?

Joyce H. Townsend

Introduction

Microfadometers of various designs have been used in museums worldwide for over 15 years to provide information on the lightfastness of materials suspected of being very light sensitive: mainly paper-based works of art such as modern prints and paper-based ephemera, textiles and other organic materials. Many instruments are closely based on the original design by Whitmore (Whitmore *et al.* 1999) and some institutions, such as Tate, have developed an instrument for a specific project (Lerwill *et al.* 2008) (Figure 1), while others have improved on the design and thereby produced unique instruments (Liang *et al.* 2011).

The method exposes a 200–600 μm diameter spot of material to approximately 6 megalux of ultraviolet (UV) and infrared-free visible light from a xenon arc lamp for 15 minutes, while the reflectance spectrum is captured every minute and processed using the CIEDE2000 colour-difference formula (CIE 2001). The Tate system uses a 600 μm diameter spot, appropriate for a hand-painted, variable surface. The results (ΔE00) are compared with ISO Blue Wool standards BW1–BW3, exposed under the same conditions and used as calibration standards. Each BW fades approximately twice as rapidly as its next highest numbered equivalent, BW1 being the most fugitive. The resulting 'blue wool equivalent (BWE)' is then used to assign individual materials to fading risk categories within which recommended display periods are proportional to fading rates (Pretzel 2008; Ford and Smith 2011).

Figure 2 shows two methods of presenting the results in which the bar chart is usually presented with the colour changes for BW1–3 included, and the microfaded areas realistically colour-coded. This format has proved popular with conservators – indeed much more popular than the raw documentation gathered during typical microfading (Figure 3). The table in Figure 2 gives unrounded numbers for the calculated colour differences and BWEs in both CIELAB1976 and CIE2000 colour spaces, as well as the colour change expressed in terms of the changes to chromaticity coordinates L*, a* and b*, and changes in chromaticity and hue. Visual examination of the bar chart format better indicates the measurement errors and estimation involved in the process of arriving at a BWE value: in general a measurement error of plus or minus half a BW is assumed. Highly textured or variegated surfaces, achievable with oil paint, might lead to even larger measurement errors.

By comparing BWE values, the least lightfast component can be identified or the relative sensitivity to light of all colours present can be compared. For paintings, when the work includes unpainted, bare canvas, this might warrant restricted display, regardless of the light sensitivity of its painted areas.

Acceptable rates of colour change vary, and a number of lighting guidelines based on

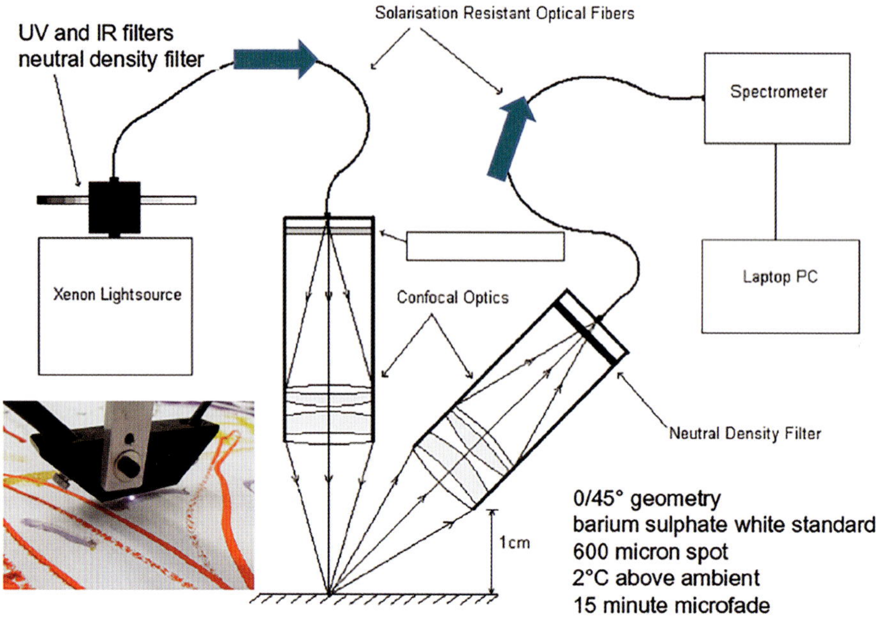

Figure 1 Schematic of a typical microfadometer, with inset colour image of the Tate microfadometer head in use for case study 2. Light scattered from the line of paint gives the false impression of a larger area being microfaded. (© Tate, 2016)

T06745 Heron	CIE76 (CIELAB)			CIE2000							
Colour	BW Range	BWE	DE76	BW Range	BWE	DE00	DL*	Da*	Db*	DC	Dh
BW1			8.3			2.3	1.8	-4.4	7.2	-7.1	-6.7
BW2			4.6			1.3	0.7	-0.8	4.6	-4.4	-2.0
BW3			0.7			0.1	0.0	-0.4	0.6	-0.7	-0.1
BW4			0.4			0.3	-0.1	0.4	0.1	-0.2	0.8
1 white priming	>BW4	>BW4	0.3	BW3-BW2	2.7	0.4	0.1	-0.1	-0.3	-0.3	0.2
2 buff 1	BW4-BW3	3.7	0.5	BW3-BW2	2.7	0.5	0.4	1.1	-4.1	1.1	-1.8
3 lilac 1	BW4-BW3	3.8	0.4	BW3-BW2	2.8	0.4	0.1	0.9	-2.6	1.9	8.5
4 buff 2	BW3-BW2	2.7	1.8	BW3-BW2	2.6	0.6	-1.8	0.4	-1.7	-0.7	-10.1
5 lilac 2	>BW4	>BW4	0.1	>BW3	>BW3	0.1	-0.2	0.7	-2.0	1.5	6.8
6 buff run	>BW4	>BW4	0.3	>BW3	>BW3	0.0	0.0	0.4	-5.2	-5.2	-1.0

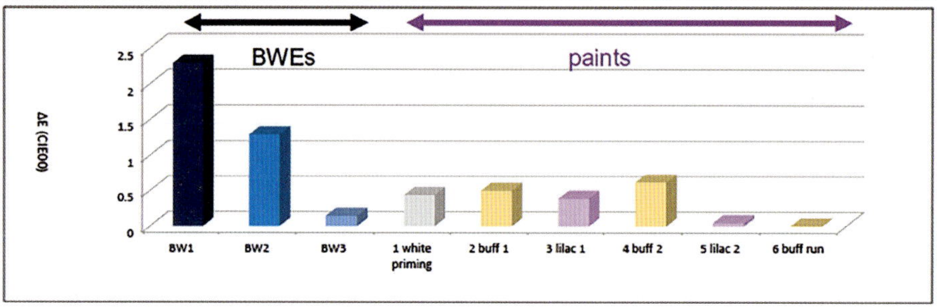

Figure 2 Microfading results for case study 2 on lilac paint discoloured to buff. (© Tate, 2016)

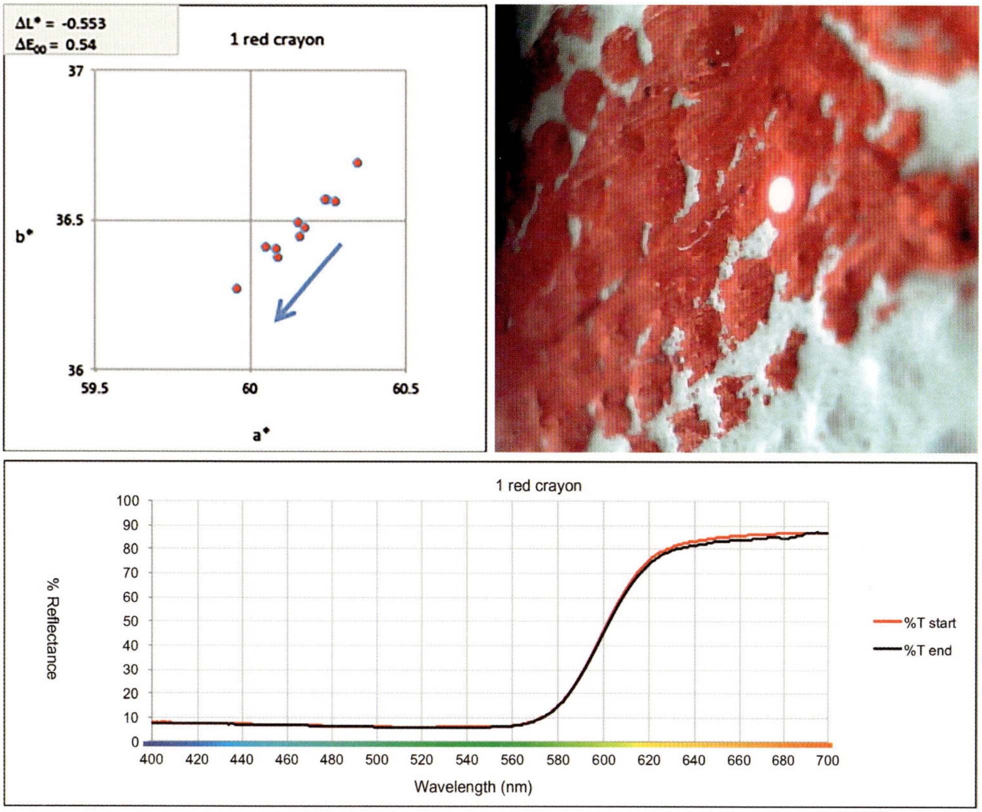

Figure 3 Raw data from microfading, case study 4. The area consisted of a red crayon-like material applied to a thin offwhite priming that does not conceal the canvas weave, and the area microfaded is one canvas thread-top in extent. It became darker, less red and less yellow after microfading. (© Tate, 2016)

fading rates of colorants compared to those of similarly exposed BWs have been published. For colorants more fugitive than about BWE3 in CIE2000, described variously as 'highly light sensitive' or having 'high responsivity' to light for museum purposes, the Victoria and Albert Museum's recommendation is 2 years display in 10 (Ashley-Smith et al. 2002), while the CIE (2004) is more conservative at 1 year in 10, both at 50 lux. Tate currently allows a maximum of 2 years display in 4 at 50–80 lux (UV-free) for works on paper when nothing is known in detail about their colorants or response to light, which corresponds to BWE3. Works found to have a BWE less than BWE3 might be recommended for display for 1 year in 4, 6 months in 4 years or even 3 months in 4 years.

There is currently an assumption at Tate that paintings that do not include bare canvas may be exposed continuously to 150–200 lux (the range being necessary to ensure even lighting of huge canvases) in acknowledgement that their constituent materials are generally much less light sensitive than watercolour paints. However, some paintings are nonetheless perceived as light sensitive and have sometimes been refused for loan even in the absence of any evidence. Interestingly, it is usually mid- and later 20th-century works that fall into this category, both for paintings and works on paper, because many artists in this period chose to use newly invented and non-traditional products whose quality control and stability would not have been optimised at the time of use. Likewise, a case could be made that some earlier 20th-century works were created by artists with equally experimental techniques and non-traditional materials. Tate's extensive building programme is leading to increased display space and predicted pressure to display iconic works almost continuously, so there is a greater need to apply microfading to a wider range of works and materials.

Because microfading involves capturing the reflectance spectrum of the area being microfaded, it potentially yields information on the material(s) being examined. It can certainly be used to compare areas suspected of having the same colorants, and thus may contribute to a technical study of the work. The nature of the colour change can also be explored and characterised as loss of yellowness, loss of chroma as a single colorant loses colour, or a more complex shift as one colour in a mixture loses colour while the other remains stable or fades more slowly.

The works discussed in this paper are all in copyright, and have been illustrated in long view only or as technical images.[1]

The difficulties of working with large paintings

The typical scenario for microfading a work on paper involves un-framing it, leaving it static on a table, and using a precision stage to move the microfadometer head over its surface. The typical microfadometer system includes a horizontal arm with a reach of around half a metre. A few rotations of the work will bring all of its surface within reach. Paper can generally be weighted at the edges with small glass blocks or fabric-covered lead weights, so movement within the sample is not an issue. Most of the paper and paint lies in plane, so focusing can be set up to one side of the object and then adjusted over the work with a fine focus control. Vibration is only an issue if the examination area is cursed with adjacent air-conditioning plant or a goods lift with a heavy-duty motor. Tate has both in the paintings studio, and a large and silent-running public lift next to the prints and drawings collection, where works on paper are most conveniently accessed. Their intermittent building-borne vibrations are only apparent after microfading. This consideration means that only the largest, therefore heaviest, examination tables with chunky legs and no wheels are suitable for microfading.

Vibration is an even greater concern when a large painting on a flexible canvas support is being microfaded. The same mitigation measures routinely applied for works that will travel need to be employed: stretcher bar linings and panel inserts should be retained in place. A stretcher bar lining planned for a floppy canvas should be fitted before microfading. Auxiliary framing devices can offer stability so the painting should have only the glazing removed. All these measures bring the obvious disadvantage of added weight: it takes a team to rotate the work on the table. It becomes easier to move the microfadometer round a static, flat painting whose paint-heavy canvas may sag down several millimetres in the gaps between the stretcher bars. Focusing is more manual, slower, and involves supporting the weight of the horizontal bar extended over the paint, some 10 mm above it (see Figure 1). The heavy stand that gives stability to the microfadometer has to be moved precisely, very close to the work and potentially all round it, moving the laptop, trailing cables and delicate optical fibres round stepwise as well. The half-metre reach of the horizontal arm means that square metres of paint surface are inaccessible when the painting is flat. Setting up takes at least as much time as the microfading itself.

Since the typical microfadometer head can be rotated 90° or more, putting the painting on an easel is a possibility or alternatively it can be leant against a wall. Minimising floor-borne vibration becomes key, since some flooring systems involve wood, are sprung, or, worst of all, are 'floated' on layers of foam for insulation. Only the most solid easels have proved to be useful for large canvas paintings, standing on load-spreading feet rather than wheels, enabling the canvas to sit truly vertical for preference, and supporting the weight of transit frames or auxiliary frames, a load of potentially 200–300 kg. The floor must also be solid since portable tables – trolleys would be worse – can transmit the vibrations from moving trolleys of artworks, packing activities in the same area, forklift trucks at work and the like.

Paint: a complex system

In principle, the interpretation of the results of microfading is fairly straightforward: provided the most light-sensitive component is of significance (a stain or a preserved sample on the reverse of a completely faded textile dye on the front might not be) then the whole work should be treated as if it has that sensitivity, even if all the other components are much more resistant to fading. This type of interpretation depends critically on the existence of a simple process for colour loss, assumed to increase linearly with increased exposure to light (known as the reciprocity effect), and not dependent on dark storage with progressive chemical reactions initiated by previous light exposure, sensitivity to an agent such as gaseous pollutants not mimicked by microfading, past treatments or chemical processes that occur only or at a different speed under intense light. Incorporating filters to reduce heating of the area being microfaded (typically to 2°C above ambient temperature) and to eliminate the UV light that the lamp may emit are obvious precautions, standard in most variants of contemporary microfadometers.

When works on paper are examined, it is usual to microfade the paper support too, if it is a visible part of the composition. It is also customary to discover, if the paper is white or off-white, that microfading makes the paper less yellow because paper forms yellow degradation products as it ages in both the light and the dark, and microfading bleaches them to other compounds. Their potential gentle bleaching during light exposure tends to be viewed as a benefit and certainly a risk worth taking. The same phenomenon occurs with paint, especially paint stored in the dark (Townsend *et al.* 2011), which means that by far the largest colour change measured during microfading is likely to be in exposed (off)white priming or in white highlights. This might not cause as much concern as actual loss of colour from a pigment.

For traditional easel paintings, the surface appearance is a consequence of thin layers of paint intentionally influenced by the colour of underlying layers, made from coloured pigments, inert extenders that are near-colourless in oil medium, and the paint medium itself, which ideally encapsulates and protects the pigments from the air. Obviously yellowed varnish would bring its own complications – by its nature it shields the paint beneath from UV and to some extent from blue light. Microfading would affect the varnish first, then, once it had bleached its yellow degradation products, it might (or might not) begin to affect the underlying colours. I am not aware of any attempts to microfade traditional, varnished paintings. For modern, unvarnished, paintings which might incorporate non-traditional paints, mixtures used in ways not advised by the manufacturer, very lean paint and/or poorly wetted pigments, the assumption that loss of colour by fading is the only relevant process must be critically examined case by case.

Artists' oil paint, acrylic paints and many house paints are designed to have much higher lightfastness than historic dyes or the organic pigments found in traditional watercolour paints. The pigments used today are often very stable. Their lightfastness is quoted by some manufacturers in their online or printed catalogues for paints, and typical figures would be BW7–8; any pigment of BW6 or less would usually be classified as having 'intermediate' stability. Bearing in mind the gain in stability of a factor of 2 for each higher BW, a pigment rated as BW8 is then approximately 32 times more stable than one rated at BW3, which is the microfading result regarded as making a work on paper 'suitable for normal

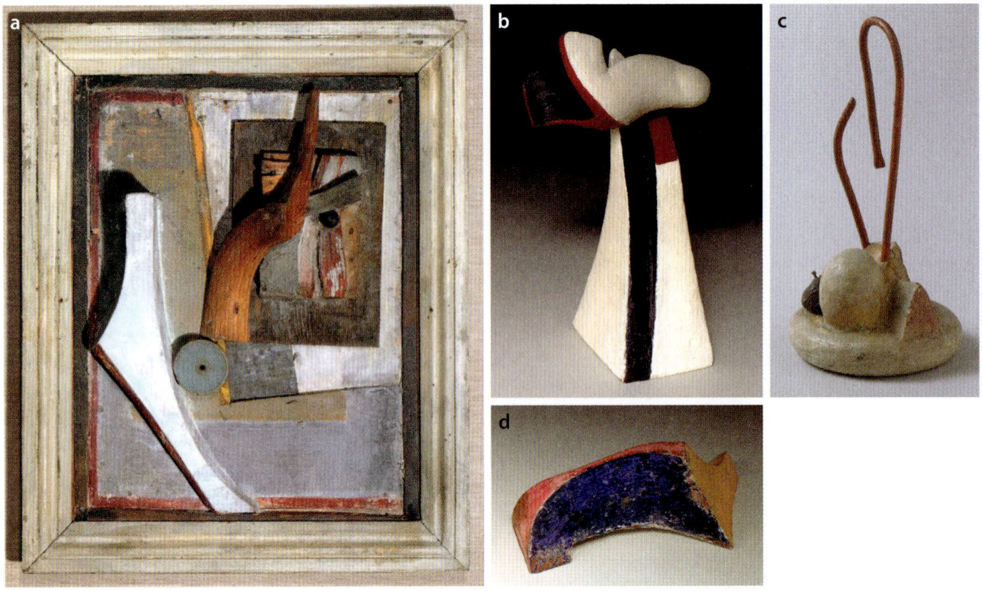

Figure 4 Late sculptures made in Britain by Kurt Schwitters. Dimensions vary from 15 to 30 cm. (a) *Relief in Relief*, T1259, *c.*1942–5, oil paint on wood and plaster. (b) *Lofty*, T05768, *c.*1945–7, painted plaster. (c) *Red Wire Sculpture*, T01259, 1944, metal, plaster and stone. (d) *Untitled (Ochre)*, T05769, *c.*1945–7, plaster. (All images © DACS, 2016 and Tate, 2016)

Tate display'. Microfading for a period long enough to 'measure BW8' would take a very long time and be impossible to achieve in practice: stray vibrations, incursions of colleagues into the workspace etc. could be expected to compromise all such lengthy measurements. In practice, only paintings suspected of having unusual sensitivity to light are microfaded. In general, what is being microfaded is a paint, not an isolated pigment, and paint formulation can affect lightfastness.

Since the essence of the process is to assess all distinct colorants in a work, the microfading of many paintings involves full technical study, preferably in advance, in order to enumerate them, and then to locate accessible areas of each on the surface.

Case study 1: Schwitters' late British sculptures

This relatively recent project (Townsend *et al.* 2014a) involved microfading combined with technical study aimed at identifying the paint medium rather than the colorants, both intended to improve the collection care of a previously rarely displayed group of works being prepared for a multi-venue exhibition. These are the surviving examples of many small works made by Kurt Schwitters when his circumstances made it challenging to source materials. He was, however, an artist who had deliberately used cheap and non-artists' materials to make a political point for most of his life. Not only were they being displayed with paper-based collages and assemblages that would be lit as sensitive works on paper (Townsend *et al.* 2013), but they would be packed for travel and thereby rendered suitable for further loans and increased light exposure in future. It was easy to see just how many paint colours and found materials had been incorporated in each (Figure 4), although the small scale of some meant that not all colours could be reached by the microfadometer head. There were no varnishes or coatings present. The comparison of medium analysis and microfading indicated that all the examples of artists' paint were stable to BWE3 (indeed, they might have been far more stable) and therefore suitable for display 2 years in 4 at 50–80 lux, while the least stable component (Hansa red, PR4, a major red colorant in three of the 12 works, and present in a mixed red in another) was less stable than BW1 and was suitable for display only 3 months in 4 years. The examination of construction using X-radiography from several sides also indicated a range of fragility, some works being nailed together securely while others were held together only by skims of plaster and paint. The combined results could be used to determine which sculptures should be offered for loan in future, as examples of this artist's last three-dimensional works.

Case study 2: light-induced discoloration rather than fading

Patrick Heron made *10-11 July 1992* (Tate T06745) in the year he dated it, using an Australian paint brand, Art Spectrum Colours, which has been the subject of recent research due to the acquired water solubility of some oil paints. It is not varnished. More than 20 years later, there are many examples of still-soft and/or poorly dried areas, especially where the artist squeezed tube paint onto the primed canvas and retained the full 3–7 mm thickness. One shade of lilac paint shows both a discoloration to buff in the thickest and most medium-rich areas, and a buff-coloured exudate running down from these areas (see the colour inset in Figure 1). A family painting, loaned for the study, showed similar issues but greater colour loss in the lilac paint; it had been exposed fairly continuously to unfiltered daylight in a domestic setting. Many of the paints in Heron's studio were given to Tate by the artist's family following his death in 1999, and in 2004 these were systematically painted out and exposed first to studio lighting and latterly on an office wall to increase their exposure to UV light. In Tate studio lighting, they developed the lilac-to-buff discoloration but not the drying issues, which suggested that light intensity, the presence of UV light and/or variations in relative humidity (RH) might all be involved in the drying process, in addition to the paint formulation of the early 1990s. These samples, now in the conservation archive (Tate Q04186) were microfaded for comparison.

Microfading with UV-free light could therefore be carried out and interpreted without too many reservations. The exposed, commercially primed canvas of the Tate painting showed less colour change than the exposed, commercially primed canvas of the family painting, supporting the idea that the latter had received greater cumulative light exposure. The stability of the lilac paint and its buff discoloration were both assessed as BW3 (Figure 2), but the change in the lilac paint was a colour loss rather than a shift towards buff: microfading did not mirror the observed colour change. The microfading study did not contribute to the study of the drying issues, but did suggest that the work should be regarded as unusually light sensitive, and exposed for '2 years in 4 at 50–80 lux', or more realistically, about 1 year in 6 at 150–200 lux when lit like other paintings.

Case study 3: comparing late 1950s and current materials

The unvarnished *Black on Maroon* (1958, Tate T0170) by Mark Rothko was damaged by graffiti ink in 2012 and its treatment, preceded by research that included a technical study and the making of a representative sample (Figure 5) with similar but necessarily current materials applied in the same sequence, has been published and discussed widely (Barker *et al.* 2014; Ormsby *et al.* 2014). The treatment benefited from an earlier study at Tate of the techniques used by Rothko for the Seagram murals (Carlyle *et al.* 2008), and concurrent research into the lighting of light-damaged Seagram murals.[2] The representative sample

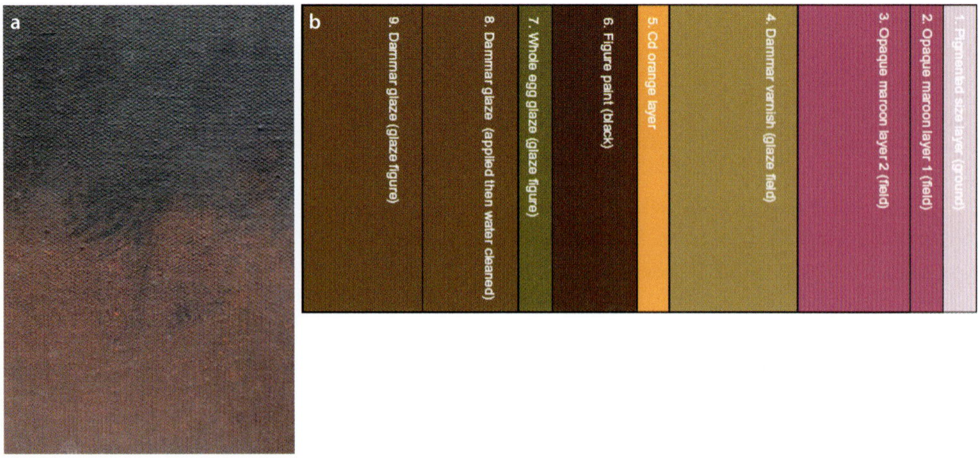

Figure 5 (a) Detail of the surface of the vandalised Rothko, and (b) schematic diagram of the representative sample (Tate Q04055) to mimic a Rothko Seagram mural. (© Tate, 2016)

was artificially aged with cycled UV-free light, modest heat and realistic fluctuations in RH, to mimic 50–55 years of natural ageing. The artist's family loaned two maroon-painted contemporary canvases (Tate loan X04205) for the project (see Figure 6), and later donated one to the Tate conservation archive (Tate Q04255). These dated to the 1950s and had received little light exposure. Tate's Seagram murals, long feared to be light sensitive, have been on near-continuous display since the 1970s, but always at lighting levels that would be deemed low even for watercolours.

Microfading was carried out on all the works and has been presented elsewhere (Townsend *et al.* 2014b). Figure 6 shows that the three Rothko canvases have broadly comparable stability, while the current materials used for the representative sample, although sourced carefully, are significantly more stable – indeed too stable for proper assessment by microfading for the usual time. This illustrates that materials in production for decades are often improved gradually, and also highlights the pitfalls of attempting to predict the behaviour of older and naturally aged materials purely from their modern 'equivalents'.

The contemporary X04205 and Q04255 both resemble the Seagram murals visually. In terms of materials, Q04255 and T01170 both have PR49 (sodium lithol red, a β-naphthol lake) in the maroon paint, while the same pigment is present in the priming of X04205, which has PV36 (reddish violet indigoid) in the paint (Kennedy *et al.* 2012; Stenger 2014). T01170 has PR49:2, calcium lithol red, in its priming. The maroon paints in X04205 and Q04255 had a lightfastness equivalent to BWE2–3, while maroon and black paints in T01170 were BWE3 or better (as was the black graffiti ink, a product called Molotow Coversall Cocktail). Given the common pigments, it appears that Rothko's thinly painted and non-covering paint layers, with visible texture on the scale of the canvas threads, lead to larger errors than usual for microfadometry, closer to plus or minus one BWE. The best estimate of the light sensitivity of these paints as used by Rothko is BWE3.

The glue-based primings, which include animal glue, two types of PR49 lithol red and synthetic ultramarine, appeared the least stable at BWE2–3. The artificially aged priming of Q04253 had, perhaps, slightly poorer lightfastness after ageing than it had originally when fresh, and the change in colour of its priming following ageing was dramatic (Figure 7): it now resembles the most altered works in the Seagram series, whose maroon paint has

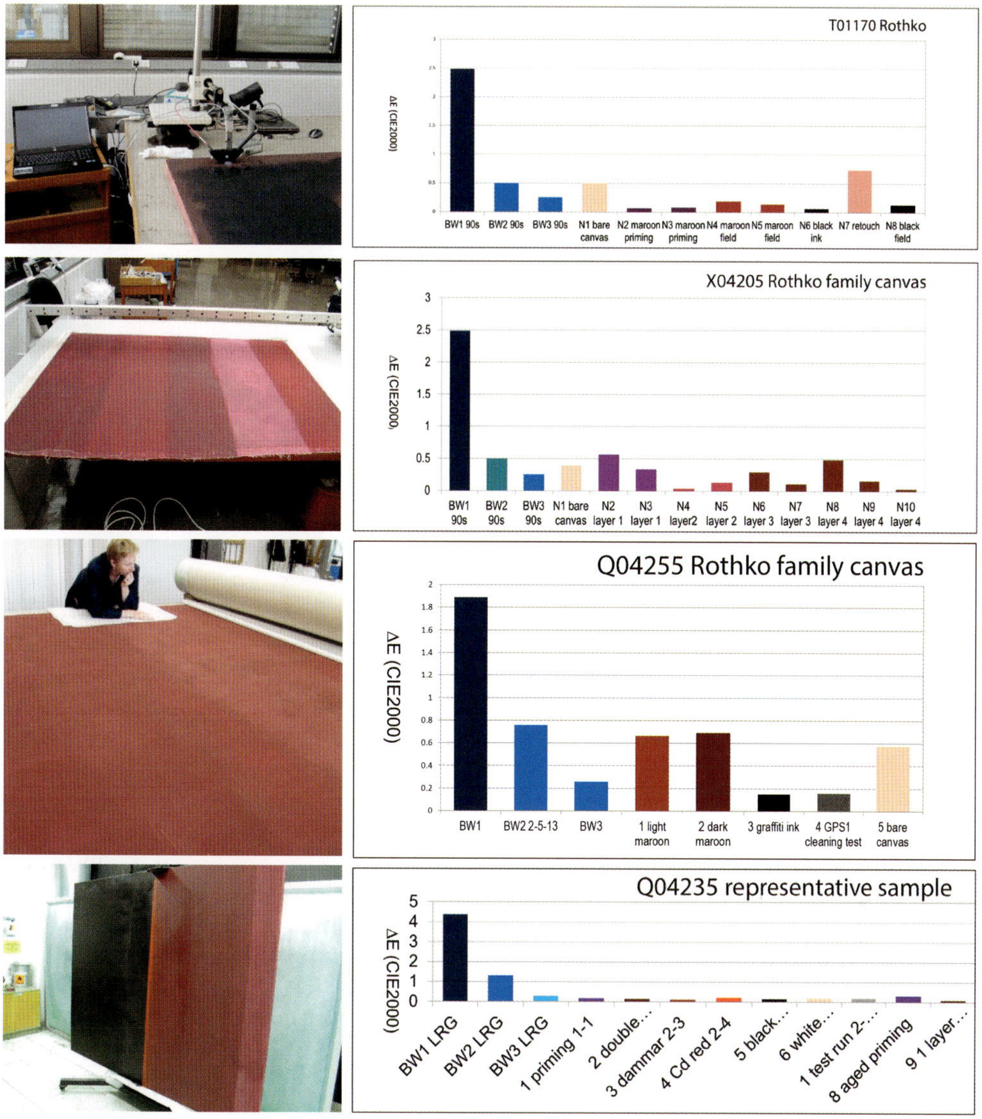

Figure 6 Microfading results for the Rothko materials and representative sample. (© Tate, 2016)

shifted further towards blue than has their maroon priming (Cohn 1988). Clearly the precise composition of the red organic pigments in each layer including the priming affects lightfastness. Since the earlier technical study (Carlyle *et al.* 2008) demonstrated that Rothko's paint layers are exceptionally thin – so much so that they follow the contours of the medium/fine cotton canvas and never fill the canvas hollows – this is not surprising. Furthermore, ongoing research suggests that Rothko's specific mixtures interact, which could not be mimicked by microfading.

The evaluation of the microfading supported the longstanding policy of regarding the Seagram murals as highly light sensitive. The recommendation that would be made, 2 years display in 4 at 50–80 lux, has been met in practice by continuous display at 30–40 lux. The nine Seagram murals (sometimes seven

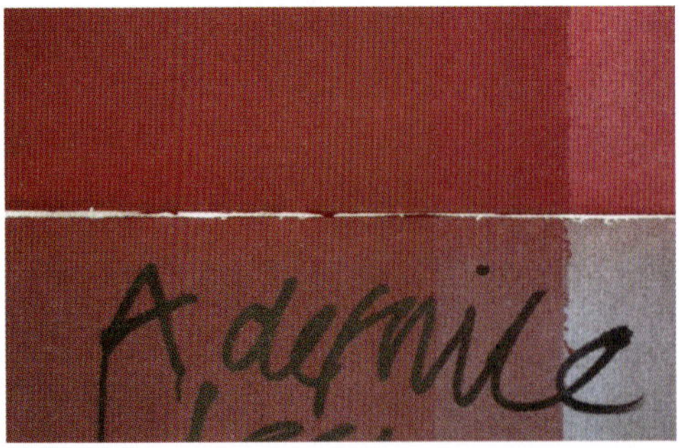

Figure 7 Portions of the representative sample Q04205, with the priming on the right and paint layers built up towards the left. Above: painted and dried. Below: after artificial ageing to mimic *c.*50 years, and application of graffiti ink. (© Tate, 2016)

or eight depending on loans) are displayed together in a single room in accordance with the artist's wishes, and it is this unusually controllable environment and visitors' tendency to linger in the room while visual adaptation to low lighting can occur that makes display at such low lighting a pleasant experience for all.

Case study 4: justifying suspected light sensitivity

Before the enormous canvases for *The Four Seasons* by Cy Twombly were packed for a loan, work schedules and workspaces were coordinated to enable all the colorants to be microfaded. The pink, purple and red areas, many highly fluorescent (which in itself promotes concerns about light sensitivity in conservation scientists), were applied as complex mixtures and overlapping layers of considerable thickness. Loan and display requests for these works have caused anxiety to many conservators since they were acquired. The four paintings are always displayed together, leaving no options for sequential loan in response to loan requests for a single canvas or display of the single canvas which would prove to be the most lightfast. It had also long been assumed that the artist had probably not used high-quality artists' materials for such large works. The four large unvarnished canvases (Figure 8) appear to share common paints and include very large expanses of unpainted, primed canvas.

Their compositions are undeniably complex and the first task was to examine them all closely in conjunction with high-resolution UV images (Figure 8), which suggested that the same colorants had been used on most of the canvases, in diluted form and/or possibly two types of medium in some cases. Close examination of the UV images clarified all the distinct colours to microfade, double-counting a few when distinction was tricky. There was evidence for existing colour loss in one (red) component. Only the stoutest and heaviest easel could be used and even then all the canvases had to be examined on their sides. This orientation meant that a strip some half a metre wide could be reached single-handedly, while art handling assistance was needed even for raising the easel, as well as for rotating the canvases. Microfading was preceded by a careful plan to access all distinct colours with the minimum movement of canvases, and was followed by a comparison of reflectance spectra to check whether all the pinks, mauve and violets were indeed distinct pigments. It was confirmed that the artist used a common palette for all four canvases, with a number of additional colours in *Spring* and *Autumn*, both rich in reds and purples. Later pigment investigation with energy-dispersive X-ray (EDX) analysis indicated that one black area and the blue, green, brown and yellow paints often included conventional artists' pigments. Given their compositions of calcium phosphate, viridian and the like, these pigments should have excellent lightfastness comparable to BWE8, although the obvious use of mixed media means that these paints might not be as stable as their colorants would imply. The reds and purples were all organic.

Microfading showed the greatest colour change for the oil- or alkyd-based priming, a very yellow white which likely acquired this yellowness during dark storage. As discussed above, this is less worrying than measured loss of chroma in a coloured pigment. The microfading results (Figure 9) show more scatter than for any of the

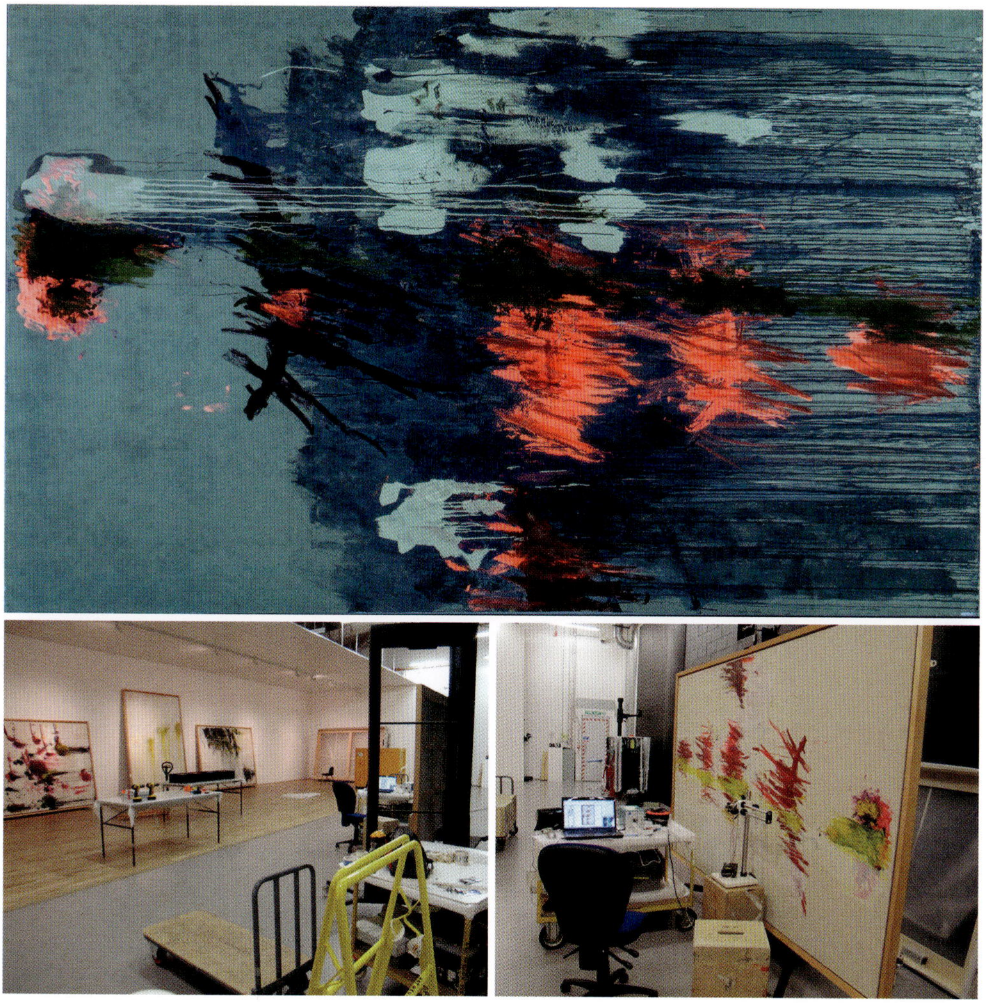

Figure 8 Lower left: Cy Twombly, *Quattro Stagione (The Four Seasons)*: T07887*Primavera*, T07888 *Estate*, T07889 *Autumno*, T07890 *Inverno*, viz. *Spring, Summer, Autumn and Winter*, occupying the 'flexible use space' at Tate Store for one week. All catalogued as acrylic paint, oil paint, crayon and graphite on canvas, all dated 1993–5. Lower right: *Spring* on the easel for microfading. Top: colour ultraviolet image of *Spring* in the same orientation. (© DACS 2015 and Tate, 2016)

previous case studies discussed. Predictably, the pinks, mauves and reds, most of which seem to be unique, had poor lightfastness (BWE2–3), although the least lightfast area was a black crayon (a mixture, not a pure black pigment, judging from its reflectance spectrum) with BWE1. The intentionally longer microfading time enabled an assessment that the colours including traditional artists' pigments were at least BWE4. The recommendation following microfading was for display for 1–2 years in 4 at 50–80 lux, or 6–12 months in 5–6 years at 150–200 lux, making these works the most highly light sensitive of an admittedly very small group of large later 20th-century paintings investigated to date.

Conclusions

These case studies show that there are occasions when microfading a complex,

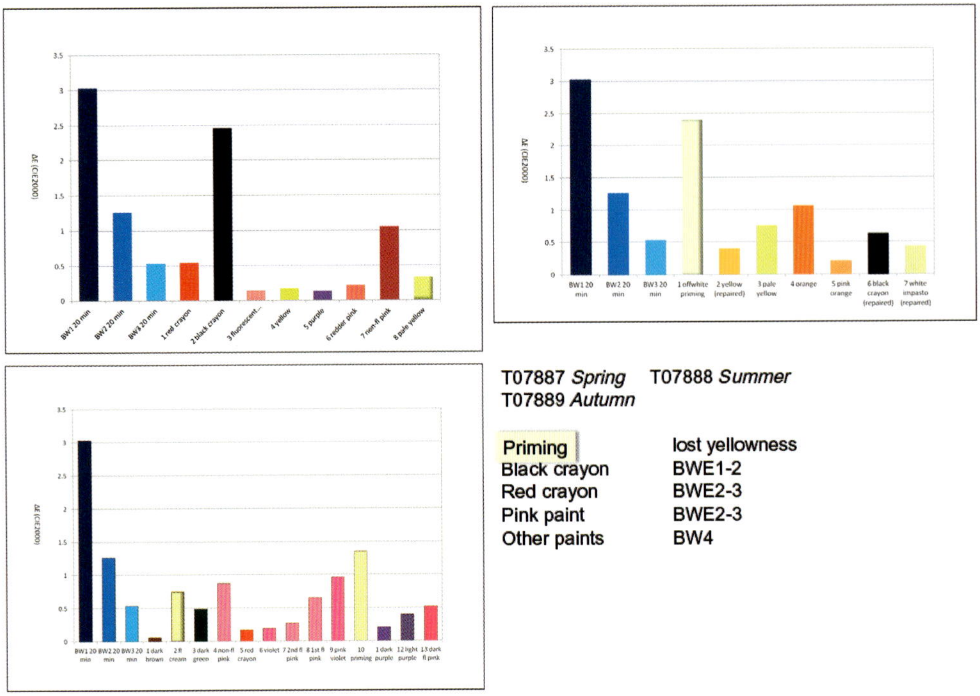

Figure 9 Microfading results for Twombly, *Spring*, *Summer* and *Autumn*. (© Tate, 2016)

unvarnished, paint surface, in conjunction with a technical study and some materials analysis, can yield very useful information for display decisions. Without the technical study and knowledge of the ageing and chemistry of oil-based paints it would be difficult to draw firm conclusions on exactly which microfading results should feed into the decision on overall sensitivity to light of a given work. Microfading a large, heavy and delicate painting on canvas is technically challenging and time-consuming, and can take many days of collective effort. It should only be undertaken when there is existing evidence of light sensitivity to prompt the investigation.

Acknowledgements

The reporting format and the Excel macros used to generate them were created by Bruce Ford, Art and Archival Pty Ltd and the National Museum of Australia, who generously shares them with the international microfading community. Thanks are due to the numerous conservators, art handlers and registrars who wrestled with the logistics of removing these large works from display and storage, transporting the microfadometer across Tate sites, and then vacating workspaces for entire working days in order to minimise vibration through the floor. Tate photographers also worked under pressure to create a full set of technical images in the few days prior to the microfading of the Twombly paintings.

Notes

1. Overall images of each work may be found online, at www.tate.org.uk, by searching for the artist's surname.
2. See the paper by Stenger *et al.*, in this vol.

References

Ashley-Smith, J., Derbyshire, A. and Pretzel, B. (2002) 'The continuing development of a practical lighting policy for works of art on paper and other object types at the Victoria and Albert Museum', in R. Vontobel (ed.), *Preprints of the ICOM-CC 13th Triennial Meeting*, vol. 1. London: James and James, 3–8.

Barker, R., Ormsby, B. and Smithen, P. (2014) 'The construction of a representative sample for Mark Rothko *Untitled (Black on Maroon)* (1958)', in J. Bridgland (ed.), *Preprints of the ICOM-CC 17th Triennial Meeting*. Paris: ICOM, 1302, 1–8.

Carlyle, C., Boon, J.J., Bustin, M. and Smithen, P. (2008) 'The substance of things', in A. Borchardt-Hume (ed.), *Rothko*. London: Tate, 75–87, 237–9.

CIE (2001) *Improvement to Industrial Colour-Difference Evaluation*. CIE publication no. 142–2001. Vienna: Central Bureau of the CIE.

CIE (2004) CIE157 *Control of Damage to Museum Objects by Optical Radiation*. Vienna: Central Bureau of the CIE.

Cohn, M.B. (ed.) (1988) *Mark Rothko's Harvard Murals*. Cambridge MA: Center for Conservation and Technical Studies, Harvard University Art Museums.

Ford, B. and Smith, N. (2011) 'Lighting guidelines and the lightfastness of Australian indigenous objects at the National Museum of Australia', in J. Bridgland (ed.), *Preprints of the ICOM-CC 16th Triennial Meeting*. Almada: Critério [CD-ROM].

Kennedy, A.R., Stewart, H., Eremin, K. and Stenger, J. (2012) 'Lithol red: a systematic structural study on salts of a sulphonated azo pigment', *Chemistry* 18: 3064–9.

Lerwill, A., Townsend, J.H., Liang, H., Thomas, J. and Hackney, S. (2008) 'A portable microfadeometer for versatile lightfastness testing', *ePreservation Science* 5:17–28.

Liang, H., Lange, R., Lucian, A., Hyndes, P., Townsend, J.H. and Hackney, S. (2011) 'Development of portable microfading spectrometers for measurement of light sensitivity of materials', in J. Bridgland (ed.), *Preprints of the ICOM-CC 16th Triennial Meeting*. Almada: Critério [CD-ROM].

Ormsby, B., Barker, R., Keefe, M., Tucker, C., Donate, F. and Smithen, P. (2014) 'The removal of graffiti ink from Mark Rothko's *Untitled (Black on Maroon)* (1958); a collaborative approach', in J. Bridgland (ed.), *Preprints of the ICOM-CC 17th Triennial Meeting*. Paris: ICOM, 1007, 1–9.

Pretzel, B. (2008) 'Now you see it, now you don't: lighting decisions for the Ardabil carpet based on the probability of visual perception and rates of fading', in J. Bridgland (ed.) *Preprints of the ICOM-CC 15th Triennial Meeting*. New Delhi: Allied Publishers Pvt Ltd, 759–65.

Stenger, J. (2014) Unpublished analytical report for Tate, 23 June 2014.

Townsend, J.H., Carlyle, L., Cho, J. and Félix Campos, M. (2011) 'The yellowing/bleaching behaviour of oil paint: further investigations into significant colour change as a response to dark storage followed by light exposure', in J. Bridgland (ed.), *Preprints of the ICOM-CC 16th Triennial Meeting*. Almada: Critério [CD-ROM].

Townsend, J.H., Ford, B. and Townshend, P. (2013) 'Who's not afraid of displaying Kurt Schwitters' collages? A microfadeometry study to determine the sensitivity to light of 5 examples', in L. MacDonald, S. Westland and S. Wuerger (eds), *AIC2013 Proceedings* 1, 35–3 [CD only].

Townsend, J.H., von Aderkas, N., Ormsby, B. and Sofer, G. (2014a) 'Kurt Schwitters' British sculptures: materials analysis and assessment of fragility', in J. Bridgland (ed.), *Preprints of the ICOM-CC 17th Triennial Meeting*. Paris: ICOM, 1011, 1–8.

Townsend, J.H., Ormsby, B., Barker, R. and Smithen, P. (2014b) 'A microfading study of Mark Rothko's materials for the Seagram murals (1958–59)', in J. Bridgland (ed.), *Preprints of the ICOM-CC 17th Triennial Meeting*. Paris: ICOM, poster abstract no. 1023.

Whitmore, P.M., Pan, X. and Baillie, C. (1999) 'Predicting the fading of objects: identification of fugitive colorants through direct non-destructive lightfastness measurements', *Journal of the American Institute for Conservation* 38: 395–409.

Author's address

Joyce H. Townsend, Senior Conservation Scientist, Tate Britain, London, UK (Joyce.townsend@tate.org.uk)

An introduction to the measurement and comparison of light sources used within cultural heritage

Joseph Padfield

Introduction

The Museum Environment (Thomson 1978) has acted as a clear guide to appropriate limits and levels for a wide range of environmental issues within museums and galleries for many years. Light and appropriate conservation lux levels are discussed in great detail, clearly explaining the appropriate preventive conservation 'dos and don'ts'. Lighting exhibitions and collections has always been a complex issue, requiring a careful balance of aesthetic presentation against the potential damaging effects of light exposure. However, in the last decade museum professionals have also been required to consider the environmental implications of the lighting solutions used. Alternative and improved sources of artificial lights needed to be explored, but this has further complicated the field, leading to new difficulties for museum professionals. But what is the difference between the actual light produced by tungsten halogen, fluorescent and LED lights?[1] Why should one type of lamp be chosen over the others and what are the preventive conservation implications of each choice? An initial examination of these differences, explaining how the light produced from different sources is measured and described, is discussed alongside how these measurements are used to create colour quality values and determine relative light exposure.

The electromagnetic spectrum, light and spectral power distribution (SPD) curves

In general, the electromagnetic spectrum[3] describes the relationships between the wavelength, frequency and energy of all electromagnetic radiation, ranging from high energy gamma rays to the much lower energy radio waves. The visible section of the electromagnetic spectrum (i.e. the section that humans can see) makes up a very small part of this complete range between the higher energy ultraviolet (UV) and the lower energy infrared (IR) light. Visible light ranges from roughly 380 to 780 nm covering the full rainbow from violet to red (see Figure 1). All the light sources used within museums and galleries, including daylight, are composed of specific mixtures of these wavelengths, potentially also including UV and IR, as well as visible light. The colour and appearance of each light source, and the objects they illuminate, are determined by this specific mixture of wavelengths, with warmer lights containing relatively more red light and cooler lights containing more blue (see Figure 2).

The measurement of the relative intensities at each wavelength of light emitted from any specific light source is referred to as its spectral power distribution (SPD) curve. SPD curves can be measured using a spectroradiometer[5]

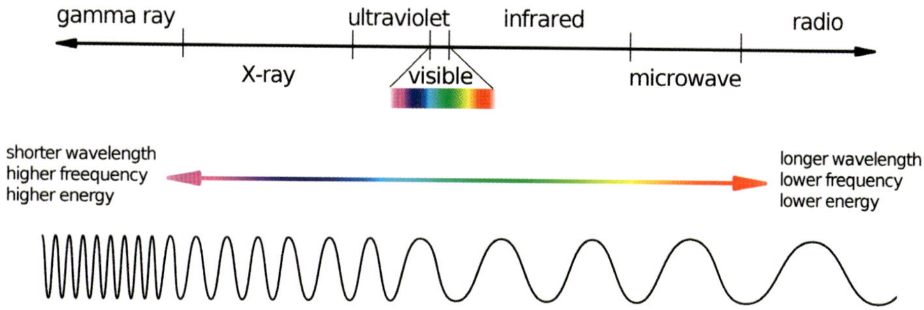

Figure 1 Graphical depiction of the electromagnetic spectrum comparing the relationship between wavelength, frequency and energy (NASA's Imagine the Universe).[2]

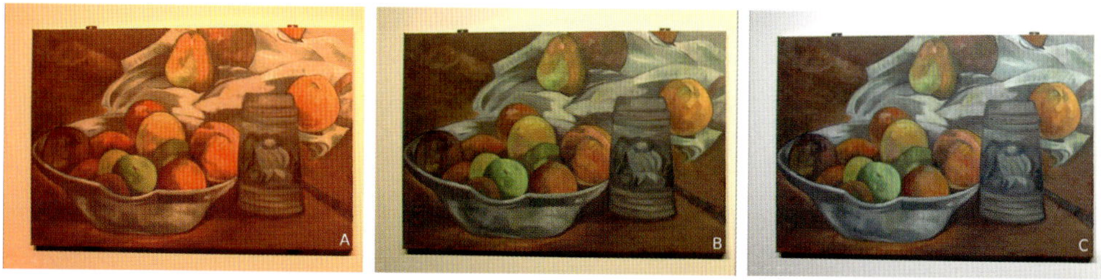

Figure 2 A contemporary copy of the central section from Paul Gauguin's *Bowl of Fruit and Tankard before a Window*,[4] 1.5 times life size, painted by Gabriella Marcaro, photographed under three different SPD curves, representing (A) traditional tungsten halogen, (B) cool white fluorescent and (C) daylight sources.

or range of different types of spectrometers[6] and they are normally displayed as graphs of wavelength against power or intensity. Four example SPD curves can be seen in Figure 3.

The curve for filtered daylight shows a characteristic, fairly even distribution of intensities for most of the wavelengths from blue to red, but drops down to zero at the very blue end, below 400 nm, before reaching the UV region, as the measurement was taken through glazing with UV filtration. The tungsten halogen curve represents a typical warm, incandescent light source[7] with a relatively high proportion of red wavelengths, steadily decreasing in intensity, moving across to the blue. The D65 fluorescent curve also displays a fairly even distribution of wavelengths, but the four very strong mercury lines typical of fluorescent bulbs[8] can also be seen, two in the blue and two in the green region, towering above the broad distribution produced by the phosphors present with the bulb. Finally, the LED curve demonstrates the typical wavelength distribution produced by many 'white' LED lamps,[9] combining a single blue LED, seen in the strong peak to the left, and one or more phosphors to create the broad bulges across the green to red regions, which combine to produce the overall white light.

As can be seen in Figure 3, SPD curves clearly show the different wavelength distributions between some of the more common types of light sources used within museums and galleries. They can also indicate the possible need for UV and possibly IR filtration by examining the spectra at the very edges of the visible region. These measured curves are also the basis for all the theoretical calculations carried out to establish the lighting values or metrics that are used to describe lights such as colour temperature (CT) and colour quality or fidelity values.

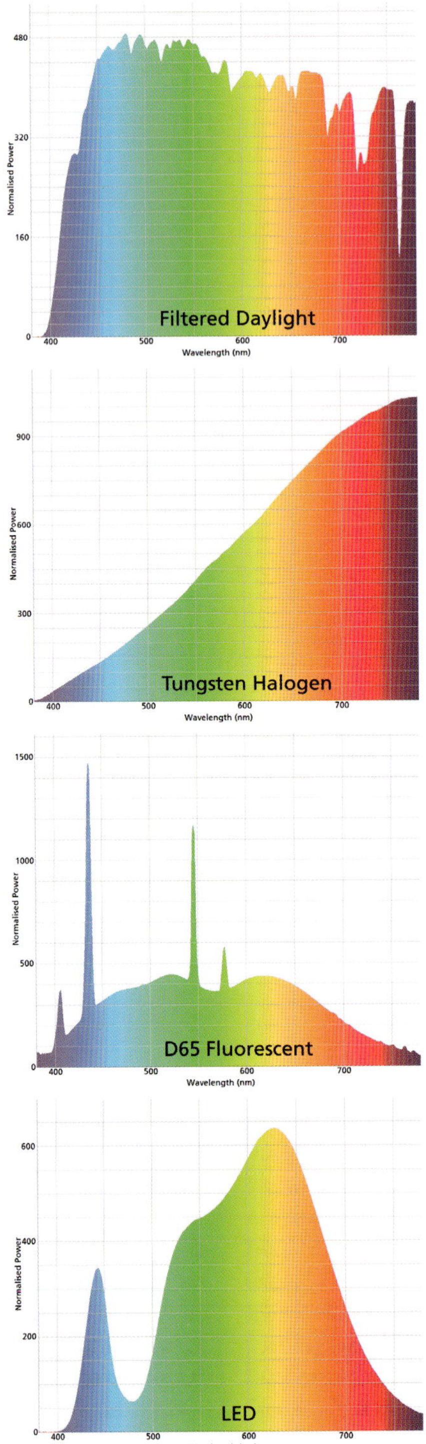

Figure 3 Four example SPD curves for four typical light sources: filtered daylight (CCT: 5595K), tungsten halogen (CCT: 2802K), D65 fluorescent (CCT: 5670K) and LED (CCT: 3034K).

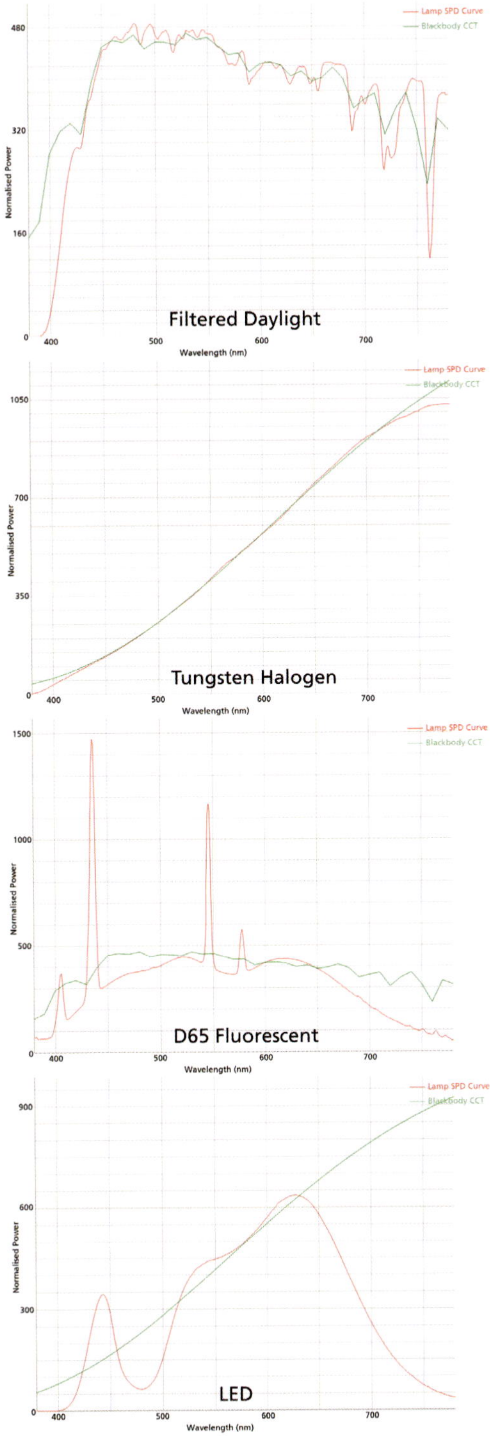

Figure 4 Four example SPD curves for four typical light sources, depicted in red: filtered daylight (CCT:5595K), tungsten halogen (CCT: 2802K), D65 fluorescent (CCT: 5670K) and LED (CCT: 3034K), each shown with a theoretical black body curve matching their CCT, depicted in green.

It is normal when measuring large numbers of the same lamp, particularly for more complex light sources such as LEDs,[10] to see some degree of variation in the SPD curves. For modern, good quality light sources, this degree of variation is often not noticeable to the human eye. For lower quality lights and more generally as different light sources age or develop faults over time, these differences can become perceptible under normal museum or gallery lighting conditions. Therefore when selecting new lights it is recommended that any manufacturer's guarantees should also consider and cover, if possible, an agreed minimum variation in the colour of a given lamp within a specific period of time, in addition to simply how long a lamp can be expected to continue to produce light.

Colour temperature (CT) and correlated colour temperature (CCT)

When an object is warmed up to a sufficient temperature it will start to emit light, for example molten metal and lava. Initially a warm red light is emitted and then, as the temperature increases, it becomes white and then finally a cooler blue-white. The SPD curve of the light produced from the object is directly determined by its temperature and this relationship – between temperature in degrees Kelvin (K) and emitted light – is described using Planck's law.[11] Ideal objects that conform perfectly to Planck's law are known as black-body radiators. The term colour temperature (CT) relates to the hue[12] of the light emitted from one of these ideal black-body radiators at specific temperatures, expressed in degrees Kelvin. The sun, and to a lesser extent flames and tungsten filaments, are relatively close approximations to these ideal black-body radiators (see Figure 4). However, many other light sources, particularly more modern ones such as fluorescents and LEDs, do not produce light in the same way, so the light they produce does not have such a closely matching CT (see Figure 4). Instead they have what is termed a correlated colour temperature (CCT), which defines the temperature of the ideal black-body radiator that most closely corresponds to its perceived colour.

The CCT of any specific light source can be calculated from its SPD curve using the CIE 1931 XYZ and xy chromaticity co-ordinates.[13] This mathematical process has been well documented (Ohno 2014)[14] and the CCT can be directly calculated, but for most applications the CCT value will automatically be calculated by the software provided with the spectroradiometer used to measure the SPD curve.

An important point to remember is that any variation in the SPD curves of a light source, as noted above, will also change the calculated CCT value. Therefore the exact CCT of any given light source will be correct for that specific light source, and some degree of variation across a batch or over time can be expected.

Colour quality: fidelity

By selecting a light source with a specific CCT the general aesthetic appearance can be determined – from a welcoming, warm light through to a cooler, possibly harsher light. However, as shown in Figure 4, different types of lamps can produce quite different distributions of light even when they have similar CCT values.[15] As a result the CCT alone does not provide any indication as to the quality of a light.

The colour quality of a light source is calculated directly from the SPD curve of a given light source and is an indication of how well the light source reveals the colours found in the objects it illuminates. This is judged against a defined reference light and scored in relation to how accurately standard observers are able to view these colours. Among the existing and proposed colour quality standards commonly used within the lighting industry, the reference light for a given test has been defined as the perfect, ideal black-body radiator with a

CT value matching the test light source's CCT value. So these particular values are only a relative not an absolute measure of colour quality. Warmer, redder light sources with low CCT values will still illuminate colours differently from a colder, bluer light source with a high CCT value, even though they both have a perfect colour quality score.

The most common colour quality or fidelity value currently used in the lighting industry is the colour rendering index (CRI) or Ra value (Houser *et al.* 2015)[16] The International Commission on Illumination (CIE)[17] CRI standard was first recommended in 1965 and then refined, updated and published as the CIE 13.2-1974 standard. The 1974 version was republished in 1995 as the CIE 13.3-1995 standard (CIE 1995); the standard did not really change but was brought up to date including adding some computer programs to assist with calculations. The CRI standard calculates the differences between the colours of the first 8 selected Munsell colour samples[18] illuminated by a test lamp and the colours produced by its matching perfect ideal black-body radiator. The calculations and colour differences are carried out using the CIE 1964 (U*, V*, W*) colour space.[19] Details of the exact calculations, including processes to account for the human ability to adjust for changes in illumination and chromatic adaptation, have been described in great detail by Houser *et al.* (2015). Calculated CRI values are determined from 0 to 100, with high quality artificial light sources, recommended for colour critical activities within museums and galleries, commonly scoring between 85 and 95.

Although still generally used within the lighting industry, the CRI standard has come under increasing criticism[20] during the last decade, particularly when it is used to describe light sources with spiky SPD curves such as fluorescent lamps or LEDs. This is because the maths used within the CRI standard was designed to work well for more traditional light sources, with SPD curves very similar to their matching CT black-body curves. Due to the method used to calculated CCT values, within the CRI standard, the values are also less reliable when used with light sources with CCT values greater than 5000 K. However, with the introduction of LEDs, although manufacturers are able to specifically design LEDs that perform well for the first 8 selected Munsell colour samples, producing a high CRI value, they might not perform as well for other colours.[21]

Table 1 A comparison of the CRI, CQS and TM-30 for the four SPD curves shown in Figure 4. The values are fairly constant for the tungsten halogen, but show a greater variation in value for other three lamps as a result of the changes in colour spaces and reference colour samples used within the three calculations.

Light	CCT (K)	CRI	CQS	TM-30 (Rf)
Filtered daylight	5595	90	97	97
Tungsten halogen	2802	97	99	99
D65 fluorescent	5670	92	94	91
LED	3034	87	85	81

These issues have led to the development of several alternatives to the CRI standard, most notably the development of the Colour Quality Scale (CQS) from 2005–6 (Davis and Ohno 2010; Ohno 2005), which utilises an improved set of 15 saturated colour samples and updates the colour difference calculations to use the newer CIE 1976 (L*, a*, b*) colour space. This was followed by CRI2012 (Smet *et al.* 2013), a proposal examining the use of even more colour samples and the more up-to-date and more uniform CAM02-UCS[22] colour space to perform the colour comparisons. Neither of these has managed to actually unseat the flawed, 40-year-old CRI standard, but the latest attempt, TM-30 (David *et al.* 2015), may be more successful. TM-30, which uses a specially selected set of 99 uniform colours distilled down from 100,000 samples and again uses the much newer CAM02-UCS colour space, has now been accepted by the American Illuminating Engineering Society (IES) as IES-TM-30 2015[23] and is in the proposal stage with the CIE.

As with the CRI calculations, describing the maths for the CQS, CRI2012 and TM-30

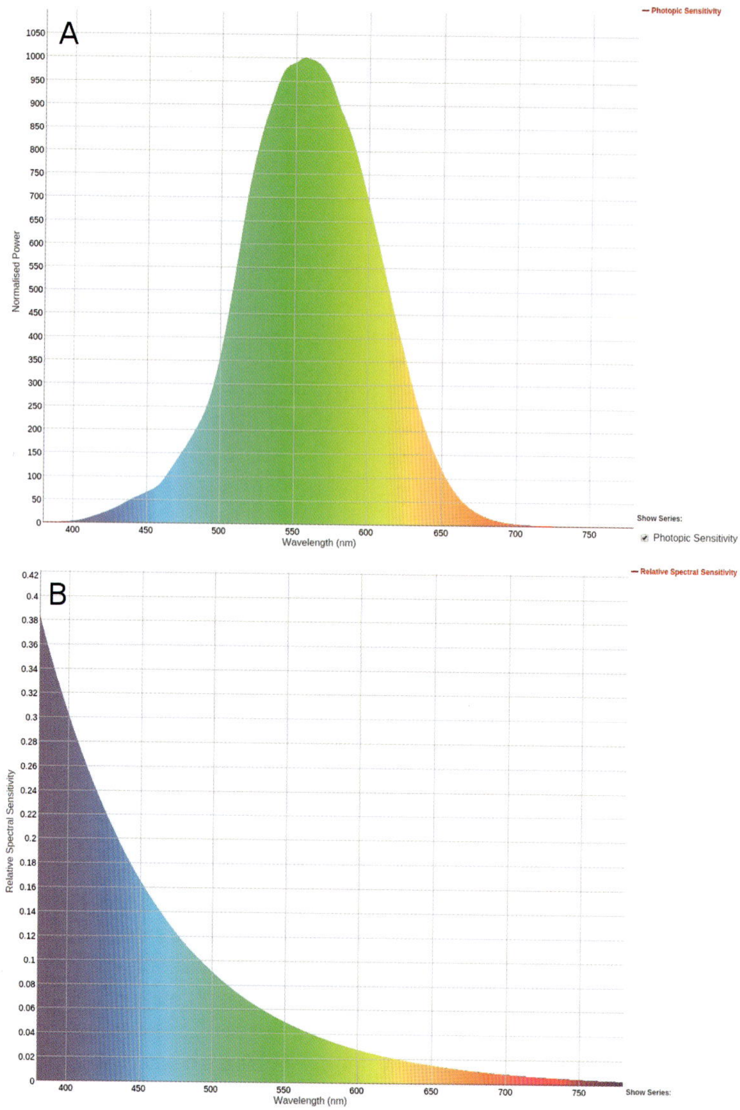

Figure 5 Two spectral sensitivity curves: (A) The human photopic sensitivity curve and (B) the Krochmann average relative spectral sensitivity curve, measured for water colours, oil paints on canvas, art paper and textiles.[25]

is beyond the scope of this paper, but further details can be found in other publications (Davis and Ojhno 2010; Ohno 2005; Smet *et al.* 2013; David *et al.* 2015). A comparison between the CRI values and two of the alternatives, CQS and TM-30, for the four SPD curves shown in Figure 4 can be seen in Table 1.

Measuring and using lux

Having defined the required aesthetic appearance – by selecting a CCT and checking how well the selected lights match that choice using colour quality values – the actual light exposure needs to be considered. How much light is actually going to hit the illuminated objects

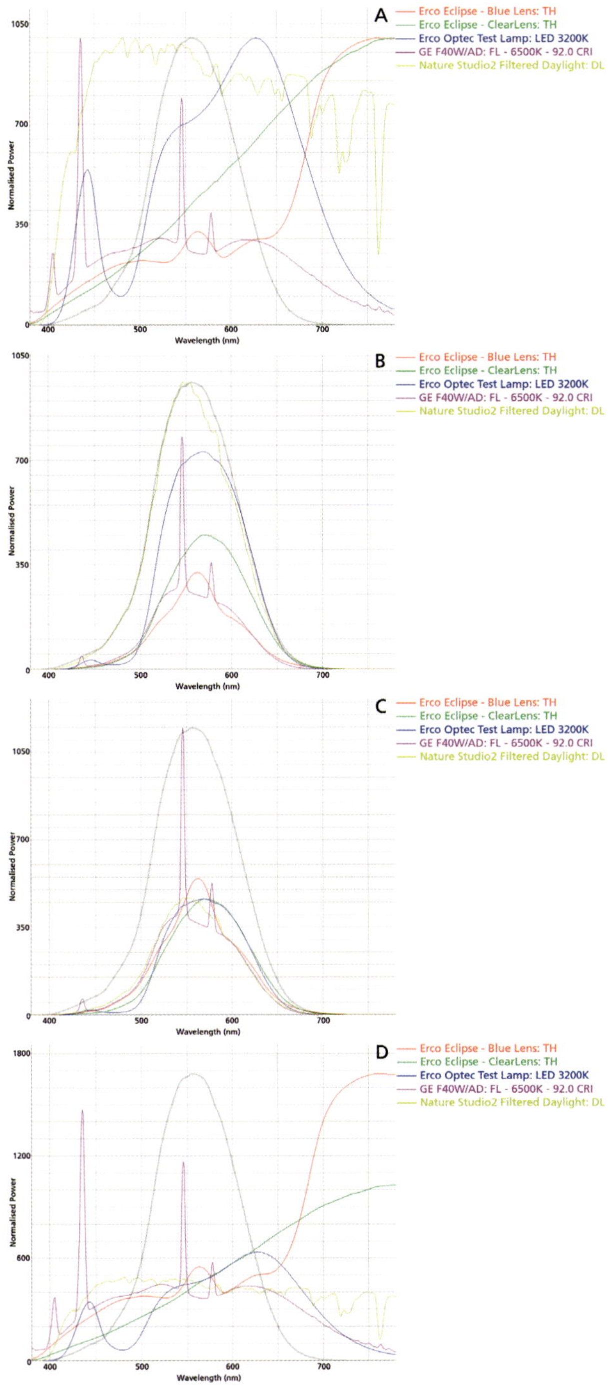

Figure 6 Four sets of SPD curves demonstrating how the human photopic sensitivity curve (depicted in grey) can be used as a filter to determine how the other curves can be theoretically normalised to a consistent lux level: (A) five example SPD curves with the human photopic sensitivity curve in grey, (B) the same five SPD curves multiplied by the human sensitivity curve, with the human photopic sensitivity curve in grey, (C) the same five SPD curves, scaled to ensure that the area under each graph is the same, and (D) five lux normalised SPD curves, with the human photopic sensitivity curve in grey.

and how does it compare? Lux[24] levels represent a measure of the brightness of a given light as perceived by the human eye. For most lux calculations, the average sensitivity of the human eye is defined by the human photopic sensitivity curve (Figure 5A), which covers a broad section of the visible spectrum centred on 556 nm. This sensitivity drops off with lower sensitivity in both the blue and red wavelengths, indicating that calculated lux levels will be dominated by the central section of a light source's SPD curve and less affected by the periphery. But unlike the human eye, works of art are sensitive to a much wider range of light, particularly higher energy blue and UV light, therefore it is helpful to examine how the SPD curves of different light sources compare with each other at the same lux levels.

Figure 6A shows five different SPD curves alongside the human photopic sensitivity curve; all are normalised so they have the same maximum peak height, which begins to allow a more direct comparison between the curves. But to compare them at the same theoretical lux level they need to be normalised again. The first step is to effectively use the human photopic sensitivity curve as a filter. Each wavelength value in the SPD curve is then multiplied by the matching wavelength value in the human photopic sensitivity curve to effectively present how each SPD would be filtered by the human photopic sensitivity curve (see Figure 6B). Next, to ensure that each of these newly scaled SPD curves represents the same level of brightness, the curves need to be scaled to ensure that the effective area under each graph is the same; the result of this process can be seen in Figure 6C. Finally these calculated scale values can then be used to scale the original curves shown in Figure 6A to produce the lux normalised curves shown in Figure 6D. Now the SPD curves can be compared at the same effective lux level, demonstrating the difference between the curves, particularly the very strong blue peaks in the fluorescent light source (depicted in purple), and to a lesser extent the LED source (depicted in blue). Figure 6D also demonstrates the much bigger difference at the red end of the visible region, particularly for the two tungsten-based light sources (depicted in red and green).

Works of art could, therefore, be exposed to a very different total amount of light energy even though the measured lux levels are the same. But, as with the sensitivity of the human eye, works of art will not be equally sensitive to all wavelengths of visible light, so further processing of the SPD curves is required.

Relative exposure percentage

When it comes to the light-induced degradation of museum objects, all light is not equal. It is generally accepted that, on average, UV and to a lesser extent blue light, can induce more degradation relative to red light due to the relatively higher energy of UV and blue light. By carefully measuring the wavelength-dependent degradation of a number of appropriate samples across the UV and visible regions of the electromagnetic spectrum, it is possible to determine this relative sensitivity to the higher energy of UV and blue light. Results published by CIE in a technical paper (Aydinli *et al.* 1991) demonstrated that, in general, the relationship between wavelength and potential damage can be expressed in the form of the following equation:

$$s(\lambda)_{dm,rel} = a * \exp(-b * \lambda)$$

where a = 1 / exp (-b * 300) and b is an average value calculated from measured reference samples. In this case a b value of 0.012 has been used, which was the value measured for water colours, oil paints on canvas, art paper and textiles. A graph of this exponential function (see Figure 5B) demonstrates that on average, across a range of materials, relative to UV light at 300 nm, objects are 0.38 times less sensitive to blue light at 380 nm and 0.005 times less sensitive to red light at 740 nm.

This relative sensitivity curve, commonly referred to as the Krochmann curve (after one

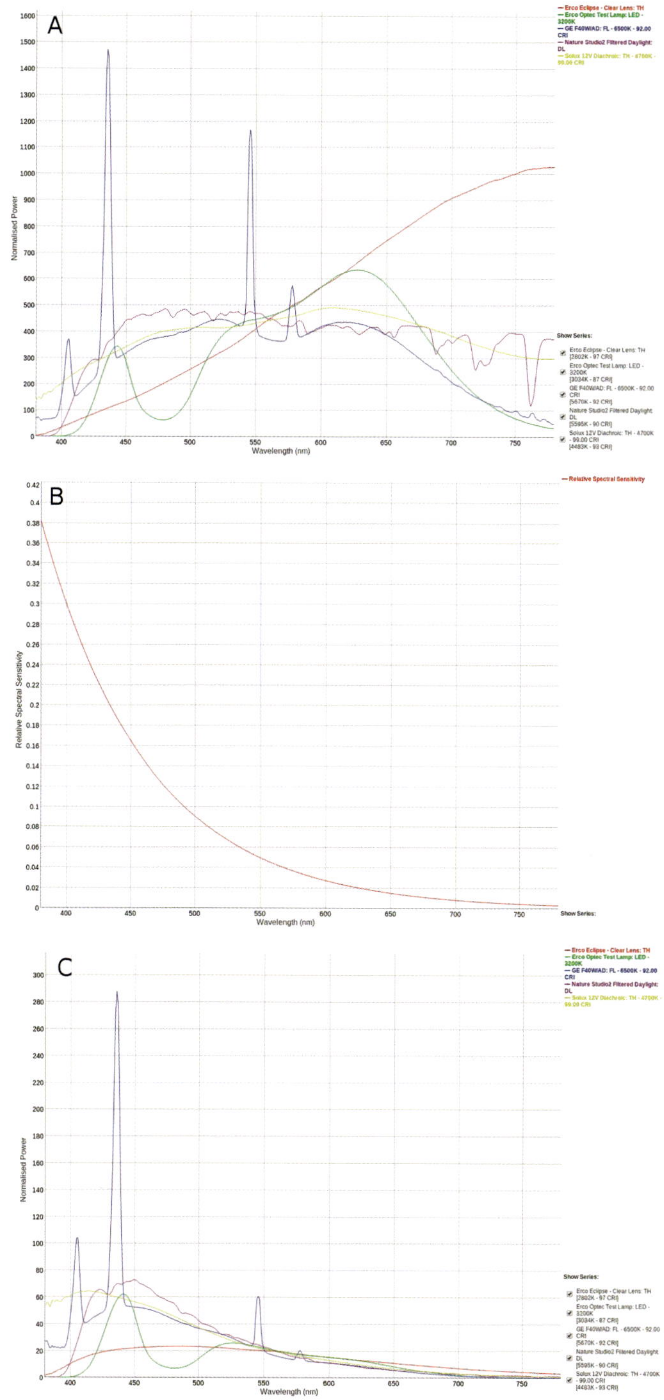

Figure 7 A series of graphs demonstrating how the Krochmann relative sensitivity curve can be used as a filter to determine how the other SPD curves can be processed to theoretically depict SPD curves normalised to match the average wavelength sensitivity of works of art: (A) five lux normalised SPD curves, (B) the Krochmann relative sensitivity curve and (C) five lux and relative sensitivity curve normalised SPD curves.

Table 2 Relative exposure percentage (RE%) values for a small range of light sources, along with the corresponding CCT and colour quality metrics.[26]

Make	Name	Type	CCT (K)	CRI	CQS	TM-30 (R_f)	RE%
GE	Vio/G4/Module/3.6W/741	LED	4121	66	81	77	127.47
GE	F40W/AD	FL	5670	92	94	91	111.75
Solux	12V Diachroic	TH	4483	93	98	98	109.46
Nature	Studio2 Filtered Daylight	DL	5595	90	97	97	100
Erco	Eclipse – Blue Lens	TH	4056	89	97	94	91.64
GE	FC8T9/CW	FL	4154	36	64	61	72.01
Erco	Eclipse – Clear Lens	TH	2802	97	99	99	63.87
GE	Par 38 80W Flood	TH	2613	97	99	99	62.02
Erco	Optec Test Lamp	LED	3034	87	85	81	58.91
GE	Standard Tungsten	TH	2313	98	98	98	57.04
CREE	LRP-38	LED	2616	91	91	86	46.23

of the authors of the Aydinli *et al.* 1991 technical paper), can then be used to further process lux normalised SPD curves to account for this wavelength-dependent relative sensitivity and obtain a clearer picture of the total amount of exposure that could be experienced by an average object from different SPD curves. Figure 7A shows a set of five lux normalised SPD curves. Each wavelength value in each of these SPD curves is then multiplied by the matching wavelength value in the relative sensitivity curve, shown in Figure 7B, to produce the lux and relative sensitivity normalised curves, shown in Figure 7C. As a result the wavelength-dependent aspects of the light exposure have been cancelled out. Therefore, the total amount of light exposure delivered from each of these SPD curves can be determined by calculating the area under each of the graphs shown in Figure 7C.

This process could then produce simple exposure totals. However, this relative sensitivity curve is not exact and represents a very average view of wavelength-dependent fading. Therefore these totals are best considered as a relative exposure amount, demonstrating how much exposure, on average, might be expected from one light source relative to another. To make the values simpler and to allow for the fact that light sources can be used at different lux levels, this relative exposure value has been expressed in the form of a percentage, calculated relative to one specific SPD curve. The National Gallery, London, is primarily illuminated using controlled, filtered daylight. Artificial light sources are selected to augment and act as a substitute for this daylight as required. Therefore, the exposure total for the filtered daylight curve, shown in Figure 3, was used as the baseline for comparison, and all of the RE% values were calculated compared to the exposure total for the measured filtered daylight (see Table 2).

This small selection of results shows that different light sources, at the same lux levels, can result in works of art experiencing very different levels of overall light exposure, with the highest RE% value, 127.47%, being almost three times the total of the smallest RE% value, 46.23%. However, by examining these RE% values it becomes clear there is no trend with a specific type of artificial light source (tungsten halogen, fluorescent or LED) always resulting in more (or less) light exposure than the others; some lamps have more, some have less. Although light sources with higher CCT values, and thus a greater proportion of blue light, generally have higher RE% values, the key to understanding the difference rests in comparing the SPD curves.

Conclusions

All visible light sources are potentially damaging to works of art, but art cannot be seen without it. Lighting works of art has always been a compromise between ensuring that all the colours within an object can be perceived and limiting overall light exposure. CCT, colour metrics such as CRI and lux values are very useful terms, but the limitations need to be borne in mind and the SPD curves of light sources should be taken into account when considering museum and gallery lighting. Different light sources, at the same lux values, can lead to very different levels of overall light exposure. Care should be taken to ensure that the lux level limits used take these potential differences into account and still act as a helpful guide, rather than an inaccurate and potentially dangerous standard.

This topic is complex, particularly when considering the maths involved. All the graphs and data provided within this paper, along with a growing amount of further data, links and information, can be found on the National Gallery SPD research website.[27] All types of lights can be 'appropriate' and 'not appropriate' for a given application in museums: the key is to take the time and select the right one.

Notes

1. Basic description of these different types of light sources can be found at: https://en.wikipedia.org/wiki/List_of_light_sources (accessed January 2016).
2. http://imagine.gsfc.nasa.gov/science/toolbox/emspectrum1.html (accessed January 2016).
3. https://en.wikipedia.org/wiki/Electromagnetic_spectrum (accessed January 2016).
4. National Gallery, London, NG6609, www.nationalgallery.org.uk/paintings/paul-gauguin-bowl-of-fruit-and-tankard-before-a-window (accessed January 2016).
5. https://en.wikipedia.org/wiki/Spectroradiometer (accessed January 2016).
6. https://en.wikipedia.org/wiki/Spectrometer (accessed January 2016).
7. https://en.wikipedia.org/wiki/Incandescent_light_bulb (accessed January 2016).
8. https://en.wikipedia.org/wiki/Mercury-vapor_lamp#Emission_line_spectrum (accessed January 2016).
9. https://en.wikipedia.org/wiki/Light-emitting_diode#White_LEDs_and_the_Illumination_breakthrough (accessed January 2016).
10. Due to the way LEDs are manufactured there can be a variation in their performance. On the production lines LEDs are separated into different quality groups, or bins, based on the degree of variation. The quality and cost of final LED-based light sources will be affected by from which quality bin the LEDs are taken. See: www.ledjournal.com/main/blogs/why-are-led-binning-ranges-important-for-led-manufacturers-and-their-customers/; http://blog.luxexcel.com/lighting/led-binning-ansi/ (accessed January 2016).
11. https://en.wikipedia.org/wiki/Planck%27s_law (accessed January 2016).
12. https://en.wikipedia.org/wiki/Hue (accessed January 2016).
13. https://en.wikipedia.org/wiki/CIE_1931_color_space (accessed January 2016).
14. https://en.wikipedia.org/wiki/Color_temperature (accessed January 2016).
15. For an example of several different SPD curves with similar CCT values see: http://research.ng-london.org.uk/scientific/spd?page=spd&ids=8,15,48,59 (accessed January 2016).
16. https://en.wikipedia.org/wiki/Color_rendering_index (accessed January 2016).
17. The International Commission on Illumination (also known as the CIE from its French title, the Commission Internationale de l'Eclairage) is devoted to worldwide cooperation and the exchange of information on all matters relating to the science and art of light and lighting, colour and vision, photobiology and image technology: http://cie.co.at/ (accessed January 2016).
18. https://en.wikipedia.org/wiki/Color_rendering_index#Test_color_samples (accessed January 2016).
19. https://en.wikipedia.org/wiki/CIE_1964_color_space (accessed January 2016).
20. https://en.wikipedia.org/wiki/Color_rendering_index#Criticism_and_resolution (accessed January 2016).
21. The selected group of Munsell samples actually includes a total of 14 colours, even

though only the first 8 are used to determine the standard CRI value. Considering the individual colour rendering index values for the some of the other patches is a common way of further testing the quality of LED, particularly the strong saturated red sample number 9. See http://www.ies.org/lda/HotTopics/LED/4.pdf (accessed January 2016).

22. Which is directly related to https://en.wikipedia.org/wiki/CIECAM02 (accessed January 2016).
23. The full Technical Memorandum document for IES-TM-30 2015 can be purchased from: www.ies.org/store/product/ies-method-for-evaluating-light-source-color-rendition-3368.cfm (accessed January 2016).
24. https://en.wikipedia.org/wiki/Lux (accessed January 2016).
25. See Aydinli *et al.* 1991.
26. Further examples of RE% values calculated for different light sources can be found at: http://research.ng-london.org.uk/scientific/spd/?page=spd-rss-table (accessed January 2016).
27. http://research.ng-london.org.uk/scientific/spd (accessed March 2016).

References

Aydinli, S., Krochmann, E., Hilbert, G.S. and Krochmann, J. (1991) 'On the deterioration of exhibited museum objects by optical radiation', CIE Technical Collection 1990, CIE 089-1991. Available at: www.cie.co.at/index.php/index.php?i_ca_id=364 (accessed January 2016).

CIE (1995) CIE 013.3-1995 *Method of Measuring and Specifying Colour Rendering Properties of Light Sources*. Available at: http://div1.cie.co.at/?i_ca_id=551&pubid=52 (accessed January 2016).

David, A., Fini, P.T., Houser, K.W., Ohno, Y., Royer, M.P., Smet, K.A.G., Wei, M. and Whitehead, L. (2015) 'Development of the IES method for evaluating the color rendition of light sources', *Optics Express* 23(12): 15888–06. DOI: 10.1364/OE.23.015888. Available at: www.osapublishing.org/oe/abstract.cfm?uri=oe-23-12-15888 (accessed January 2016).

Davis, W. and Ohno, Y. (2010) 'Color quality scale', *Optical Engineering* 49(3), 0336022. Available at: http://opticalengineering.spiedigitallibrary.org/article.aspx?articleid=1096282 (accessed January 2016).

Houser, K., Mossman, M., Smet, K. and Whitehead, L. (2015) 'Tutorial: Color rendering and its applications in lighting', *LEUKOS* 12(1,2):7–26. Available at: http://dx.doi.org/10.1080/15502724.2014.989802 (accessed January 2016).

Ohno, Y. (2005) 'Spectral design considerations for white LED color rendering', *Optical Engineering* 44(11), 111302. Available at: www.nist.gov/manuscript-publication-search.cfm?pub_id=841839 (accessed January 2016).

Ohno Y. (2014) 'Practical use and calculations of CCT and Duv', *LEUKOS* 10(1): 47–55. Available at: http://dx.doi.org/10.1080/15502724.2014.839020 (accessed January 2016).

Smet, K.A.G., Schanda, J., Whitehead, L. and Luo, M.R. (2013) 'CRI2012: a proposal for updating the CIE color rendering index', *Lighting Research & Technology* 45(6): 689–709. Available at: http://lrt.sagepub.com/content/45/6/689 (accessed January 2016).

Thomson, G. (1978) *The Museum Environment*. London: Butterworth Heinemann

Author's address

Joseph Padfield, Conservation Scientist, Scientific Department, The National Gallery, London, UK (joseph.padfield@ng-london.org.uk)

Lost purples: detection, composition and reconstruction. Rediscovering the appearance and contemporary resonance of purple paint in Tudor and Stuart portraits

Caroline Rae and Libby Sheldon

Introduction

The rare and judicious use of purple by portrait painters in Tudor and Stuart Britain was controlled not only by sumptuary laws and the sitter's social position, but by the difficulty of creating a purple hue with the pigments available in Northern Europe. Scientific examinations of paintings have shown smalt and red lake to be the most common mixture for purple on portraits, a practice confirmed by documentary sources such as Henry Peacham's advice to young artists that 'for purple velvet, take oyle smalt, and temper it with lake, halfe lake, halfe smalt; then take white lead and order it as bright or as fad as you list' (Peacham 1906: 133). However, both these colours have problems in terms of longevity and recent research has revealed the presence of many more purples, made with this combination of smalt and red lake, than are now visible on the surface.

Findings in the last few years of complete loss or change of an original purple colour have been observed on various portraits (during the 'Making Art in Tudor Britain' project at the National Portrait Gallery, London). Purple in Tudor portraits may have been much less rare than was previously thought. The mineral fluorite, for example, has now been found in a number of portraits through close scrutiny and analysis. This particular research, however, hopes to understand and perhaps reconstruct the original appearance of the distinctive smalt and crimson lake mixture, and to discover what function such lost purples had performed and what significance they represented.

Significance of purple

Since classical times, purple fabrics have been associated with the ruling classes and hence subject to legislation. In England, 18 sumptuary laws were passed between 1337 (under the reign of Edward III) and 1603 (the accession of James I). Four sumptuary laws (1510, 1514, 1515 and 1533) were passed during the reign of Henry VIII (Hayward 2009: 17). These laws directly linked social status with the quality of materials. In his 1510 law, Henry clearly stated that purple fabrics should be associated with royalty and imposed controls on the wearing of silks, luxury fabrics woven with metal threads

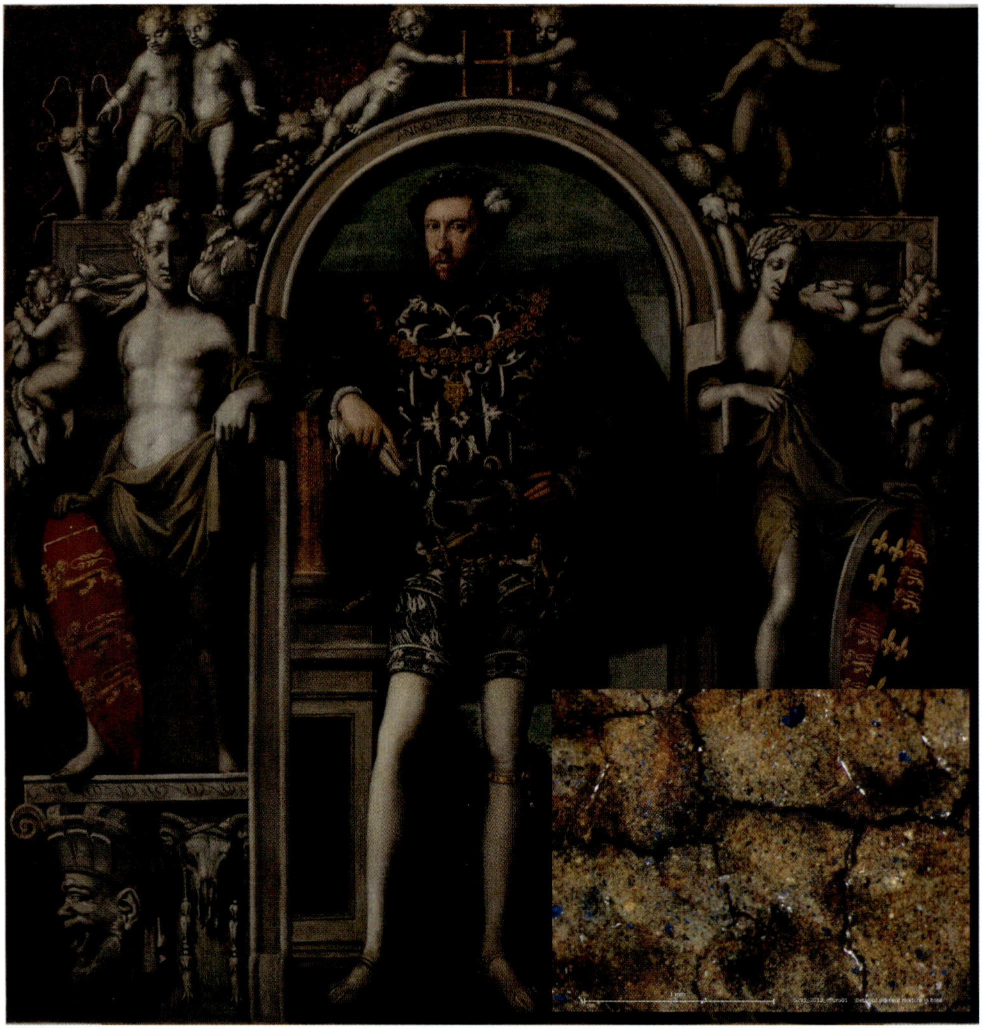

Figure 1 Unknown artist, *Henry Howard, Earl of Surrey*, c.1546, oil on canvas, 221.6 × 219.7 cm, National Portrait Gallery, NPG 5921. Inset: microphotograph of discoloured purple showing traces of smalt and crimson lake.

and furs such as sable (Hayward 2009: 17). However, luxurious purple fabrics were not restricted to royal personages alone: the king could also licence those of his choice to wear them (Bolland 2015: 354). To contemporary audiences conversant with such hierarchies of dress therefore, the rich, purple fabrics included in these portraits would have immediately conferred upon the sitter royal or elite status and the depiction of purple dress would have been a significant choice for both patron and artist.

The significance of purple in two Henrician portraits

One of the earliest portraits in which an important purple element was discoloured is *Henry Howard, Earl of Surrey*, 1546, NPG 5291 (Figure 1). Believed to have been created for Surrey by an unknown Italian artist, this portrait stands out in the context of portraits produced in England at the time due to its scale and grandiose composition featuring a *grisaille* archway which frames the sitter in

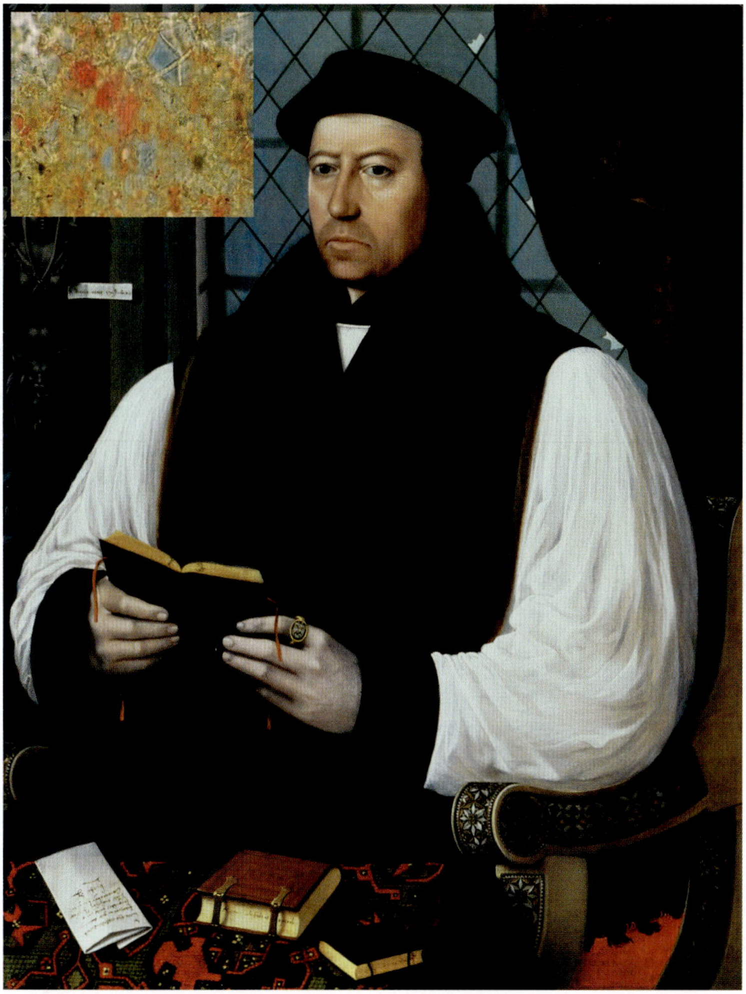

Figure 2 Gerlach Flicke, *Thomas Cranmer*, 1545, oil on panel, 98.4 × 76.2 cm, National Portrait Gallery, NPG 535. Inset: microscopic dispersion of paint from brown curtain, photographed in plane polarised light at ×1000, showing smalt and red lake particles.

a dramatic fashion (Bolland 2015: 354). At the time of creation, the painting would have made an even greater impact on the viewer. Surrey's doublet and hose were originally a sumptuous purple, created using smalt, azurite and red lake (Bolland 2015: 354). Due to the discoloration of smalt in the paint layers, the costume is now browned in appearance. Surrey was gifted such a purple jacket and doublet, threaded through with gold and silver, by Henry VIII in 1540 (Bolland 2015: 354). By having himself depicted in this costume, Surrey therefore demonstrated his closeness to the king and provided a visual assertion of his claims to power and nobility. Surrey may have overreached himself, however, as a year later he was executed by the king on a charge of treason, regal imagery in a portrait being a contributing factor (Bolland 2015: 352).

Another favoured courtier of Henry VIII, Archbishop Thomas Cranmer, could surely expect permission to include the symbolic royal purple in his portrait. It was this supposition which raised the possibility that the paint

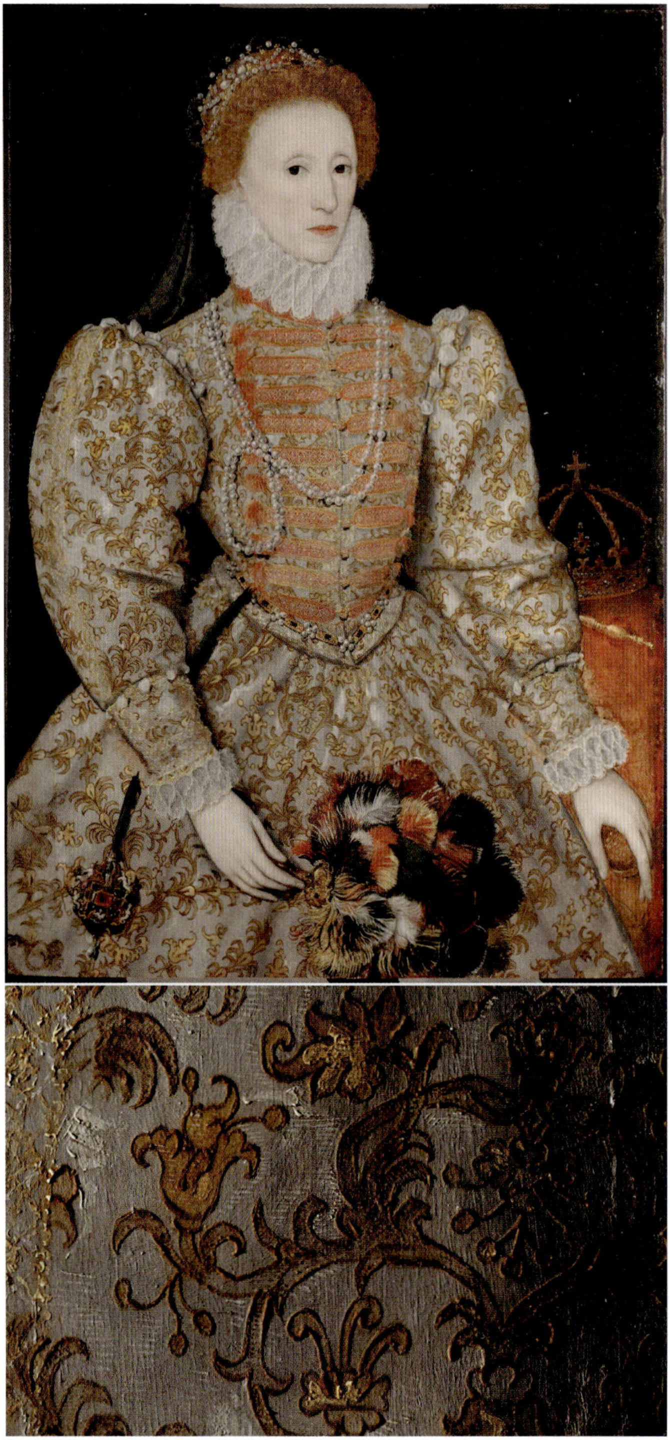

Figure 3 Unknown artist, *Elizabeth I*, 'Darnley' portrait, *c.*1575, oil on panel, 113.3 × 78.9 cm, National Portrait Gallery, NPG 2082. Inset: detail of embroidery on dress, photographed in raking light.

of the dull brown curtain behind the sitter had once been a mauve or purple colour and led to its examination. Confirmed by polarised light microscopy (PLM) of a sample, the brown's main constituents were identified as smalt and a warm red lake (Figure 2). Only the larger particles of smalt have retained some of their original royal blue colour, and it is also in the larger particles of lake that the original red hue has survived, perhaps where mostly hidden from light in the thicker passages of the paint. Simultaneous darkening of the oil medium has contributed to the present brown hue, however the remnants of both colours show the likely chromatic strength of the original pigments and intensity of the mixed purple. The care with which colours were chosen by Gerlach Flicke for this important portrait is evident from a written notation of the German word for red, *rot* (discovered by infrared reflectography),[1] which had been inscribed underneath the paint of the carpet patterns.

Undoubtedly, the most powerful and effective combination of pigments to achieve a satisfactory purple would have involved the blue ultramarine rather than smalt, but the hugely inflated cost of ultramarine in 16th-century Northern Europe would have inhibited such an extravagant and hidden employment of lapis lazuli (Sheldon 2014: 133, n. 24). Instead, smalt, a good coloristic match when first applied, could achieve an immediate, glorious colour, and would almost certainly have retained its hue for the shortened lives of both these sitters.

The manner in which smalt in oil decays – both its tendency to lose its colour by the migration of potassium into the surrounding medium and the latter's simultaneous change to a yellow-brown – has been investigated and described in detail by Spring *et al* (2005). Similarly, the loss of colour from red organic pigments through exposure and degradation by light is well understood and frequently noted (Saunders and Kirby 1994). It is the combination of these two colours with their different degrees of change which makes it particularly difficult to assess their original colour on paintings.

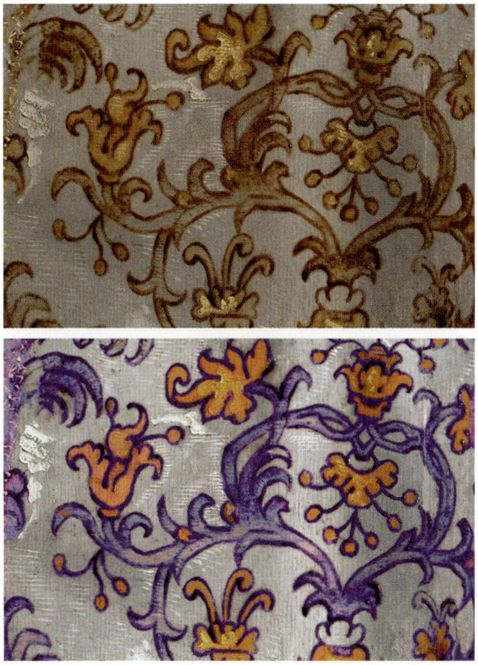

Figure 4 (top) *Elizabeth I* 'Darnley' portrait, detail of the embroidery on the dress in ordinary light; (bottom) digital reconstruction of the original colour of the embroidery in the same area.

Elizabeth I, 'Darnley' portrait, c.1577

Loss of both colours has wrought another surprising and significant colour change to the embroidery of Elizabeth I's costume, where original raised patterns of purple surrounding the inner yellow colour have discoloured to a dull orange. The presence of both large quantities of mostly colourless smalt and remnants of a crimson lake at first suggested that the prime purpose of the smalt was to give a bulked effect to the paint, in imitation of raised pink or red embroidery (Figure 3). However, scanning electron microscopy-energy dispersive X-ray (SEM-EDX) analysis (Spring *et al.* 2012) confirmed the original strength of the blue smalt, while red under the rebate on the same painting suggested the chromatic force of the red lake component. The powerful combination of the red and blue indicated that the raised

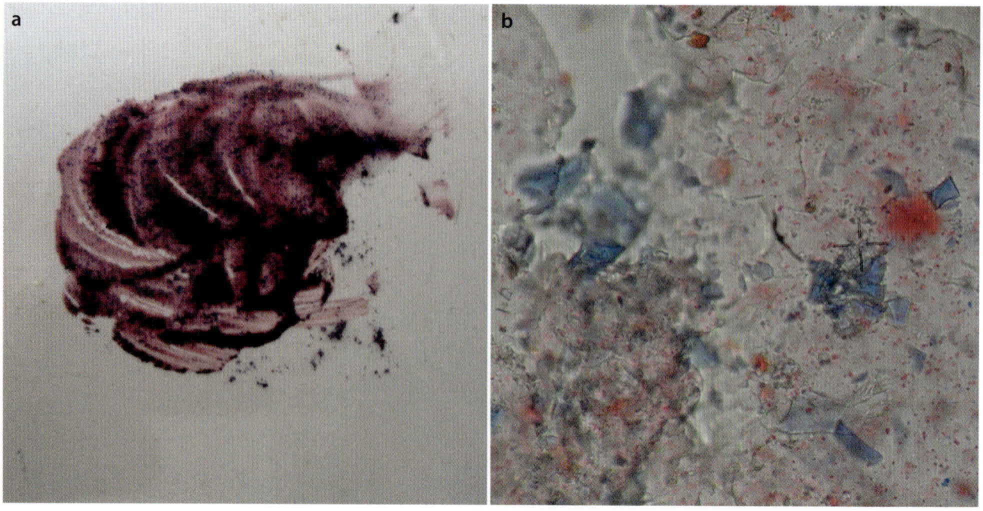

Figure 5 Reconstructed purple using smalt, red lake and lead white: (a) mixture painted out; (b) microscopic dispersion of paint taken from paint-out, photographed in plane polarised light at ×1000 to show the proportionate mixture.

embroidery was indeed a purple. An indication of the startling effect of this original colouring can be seen in the digital reconstruction (Figure 4).

Was purple a more common colour than is generally supposed in portraits of Elizabeth I? For example, a casual examination of the ribbons of another portrait of Elizabeth I (the so-called 'Armada' portrait, NPG 541, unknown artist) suggested that alternating pale pink bows on her costume, which contained colourless smalt as well as faded crimson lake, might once have been purple.

Ongoing research and reconstructions

Attempts to gauge the original colour of purple mixtures have been reliant not only on scientific analysis and documentary evidence but also on physical experiments in reconstruction. The feasibility of reconstructing these purples has been explored, attempting to match the type and proportions of pigments found in paintings through PLM. Employing, as far as possible, historically accurate pigments of smalt, red lakes and lead white, paint mixtures have been produced and painted out to give varying shades of purples. These paint-outs have then been sampled as dispersions set on slides for microscopic comparison with previously gathered dispersions of purple from 16th-century paint (Figure 5).

There are of course many problems with such experiments. Although proportions of the three pigments could be assessed to some degree from the 16th-century mixture, the resulting colour of the new mixtures relied too heavily on a visual judgement of the type of red lake employed. In the experiments to date, a strong purple has been difficult to achieve. However, this research is ongoing and hopes to add other calculations and approaches to the reconstruction research.

Conclusions

Finding traces of lost purples in just a few paintings examined at the National Portrait Gallery, London, suggests that purple may have been a much more commonly used colour than is readily apparent to the modern eye. Used in small but important touches on the ribbons of the 'Armada' portrait of Elizabeth I, for example, or in the overall costume of Henry Howard, the original impact of the purple must have been significant. Future research may provide an even clearer idea both of the spread of its usage as a colour and the appearance of the original purple.

Note

1. Infrared reflectography (IRR) carried out by Tager Stoner Richardson as part of the 'Making Art in Tudor Britain' project, hosted by the National Portrait Gallery, London. For more information see www.npg.org.uk/research/programmes/making-art-in-tudor-britain (accessed 21 March 2016).

References

Bolland, C. (2015) '"*SAT SUPER EST*": a portrait of Henry Howard, Earl of Surrey', in T. Cooper, A. Burnstock, M. Howard and E. Town (eds), *Painting in Britain 1500–1630: Production, Influences and Patronage*. Oxford: British Academy and Oxford University Press, 292–394, esp. 352–61.

Hayward, M. (2009) *Rich Apparel: Clothing and the Law in Henry VIII's England*. Aldershot: Ashgate Press.

Peacham, H. (1906) *Peacham's Compleat Gentleman, 1634: with an Introduction by G. S. Gordon*. Oxford: Clarendon Press

Saunders, D. and Kirby, J. (1994) 'Light-induced colour changes in red and yellow lake pigments', *National Gallery Technical Bulletin* 15: 79–97.

Sheldon, L. (2015) 'Palette, practice and purpose: pigments and their employment by native and Anglo-Netherlandish artists in Tudor and Jacobean painting', in T. Cooper, A. Burnstock, M. Howard and E. Town (eds), *Painting in Britain 1500–1630: Production, Influences and Patronage*. Oxford: British Academy and Oxford University Press, 128–37.

Spring, M., Higgitt, C. and Saunders, D. (2005) 'Investigation of pigment–medium interaction processes in oil paint containing degraded smalt', *National Gallery Technical Bulletin* 26: 56–70.

Spring, M., Kugler, V. and Bean, S. (2012) 'Quantitative energy dispersive X-ray analysis of the blue pigment smalt in the variable pressure scanning electron microscope', in C. Meeks, C. Cartwright, A. Meek and A. Mongiatti (eds), *Historical Technology, Materials and Conservation: SEM and Microanalysis*. London: Archetype Publications, 114–22.

Author's addresses

- Caroline Rae, PhD Candidate, Courtauld Institute of Art and the National Portrait Gallery, London, UK (carolinejrae@hotmail.com)
- Libby Sheldon, Honorary Research Associate, History of Art, UCL, London, UK (libbysheldon@hotmail.com)

Choosing colour: the reconstruction of green textiles in *The National Synod* by Pouwel Weyts

Johanneke Verhave and Lidwien Speleers

Introduction

The visual impression of a painting can change greatly as a result of paint discoloration as in the case of the painting *The National Synod* from the Dordrechts Museum (the Netherlands) in which a green glaze has turned brown. In the past, restorers of the *Synod* painting have approached this particular colour change in different ways. This paper discusses these attitudes and clarifies our choice for a reconstruction of the original colour.[1]

The National Synod of Dordrecht

In the spring of 2015 a new museum called Het Hof van Nederland (The Court of the Netherlands) opened in Dordrecht. The museum, a subdivision of the Dordrechts Museum, highlights the history of the city and its role in forming the Dutch Republic. One of the topics emphasised is the history of religion in the Netherlands, with special attention paid to the ideological conflict among the Protestants in the 16th and 17th centuries. The more traditional Calvinists clashed with the Arminians on the subject of pre-destination; a grim dispute that almost led the country into civil war. To settle the dispute a synod was held of Dutch and foreign delegates in Dordrecht in 1618–19. To commemorate the event, the city council commissioned a large painting showing the delegates and some visitors at one of the many sessions of the synod in the Kloveniersdoelen (see Figure 1) and for centuries the painting adorned the city hall of Dordrecht. After a full restoration it is now on display in the new museum.

The painting is unsigned and is ascribed to the unknown painter Pouwel Weyts the Younger (d.1629) on the basis of a document that records the payment of 84 (Dutch) pounds to Weyts for the delivery of the painting in 1621.[2] Weyts's father, Pouwel Weyts the Elder (d.1618), was a painter in Dordrecht and may have trained his son, although no other paintings can be attributed to Pouwel Weyts the Younger with certainty. The latter moved to Delft in 1620, where he was enrolled in the St Luke Guild as a painter, but archival documents refer mainly to his work as an art dealer (Marijnissen *et al.* 1992: no. 77, 279).

Painting technique of *The National Synod*

The painting is not the first visual record of the synod – from the outset engravers and publishers documented the assembly in prints. François Schillemans was commissioned to make an

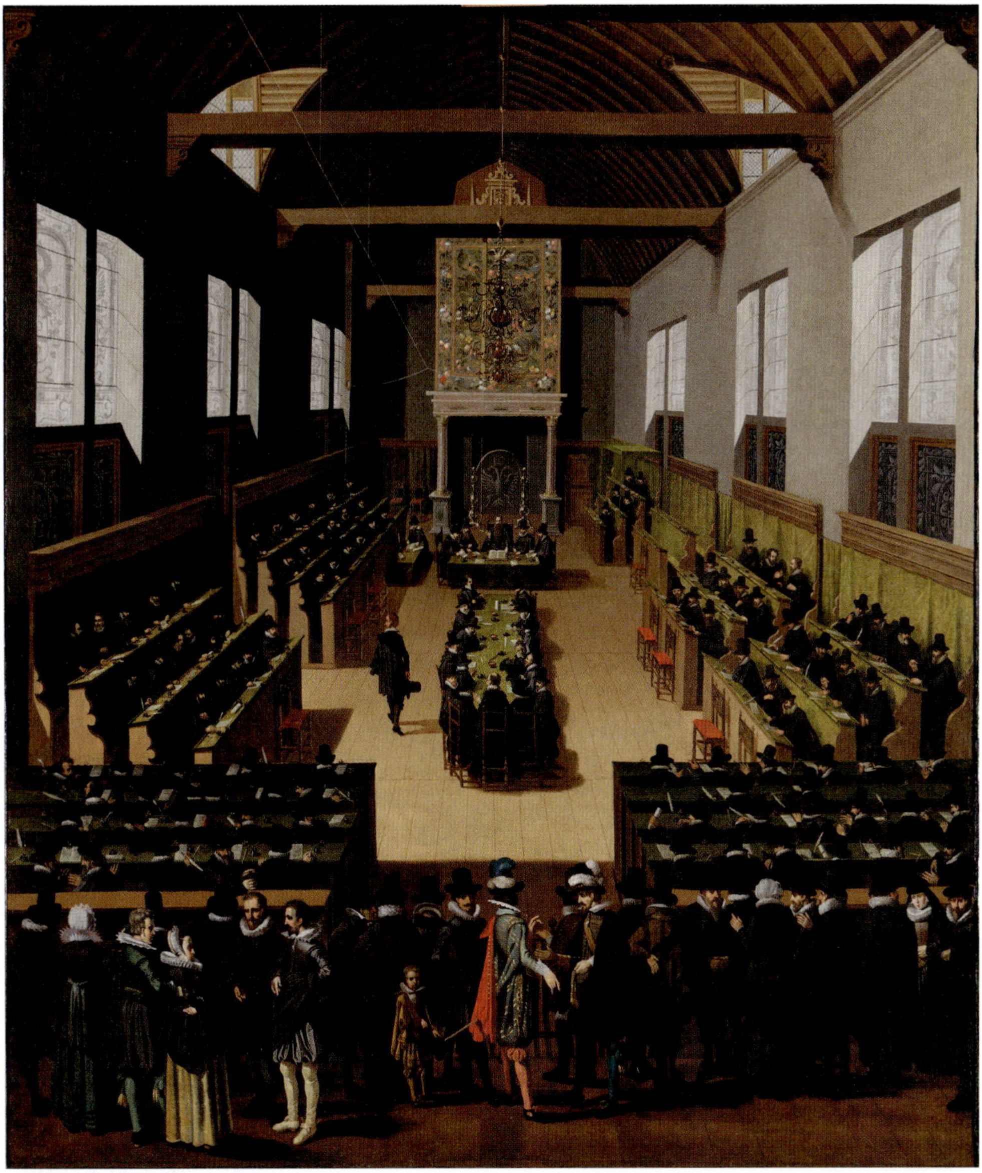

Figure 1 *The National Synod*, Pouwels Weyts the Younger (attr.), 1621, oil on canvas, 199.7 × 174.9 cm, Dordrechts Museum (DM/975/504): after restoration.

engraving which, among other things, was used as a parting gift for the delegates. Weyts based his painting on Schillemans's print.

The painting has a linen support consisting of two pieces of fabric joined by a horizontal seam running just below the central table where Johannes Bogerman, the chairman of the meeting, is seated. The original tacking edges were removed during a previous lining. The ground layer on the support is brick red in colour, as can been seen in areas of damage and open cracks.[3] No evidence of underdrawings used for the transfer of the composition was observed.

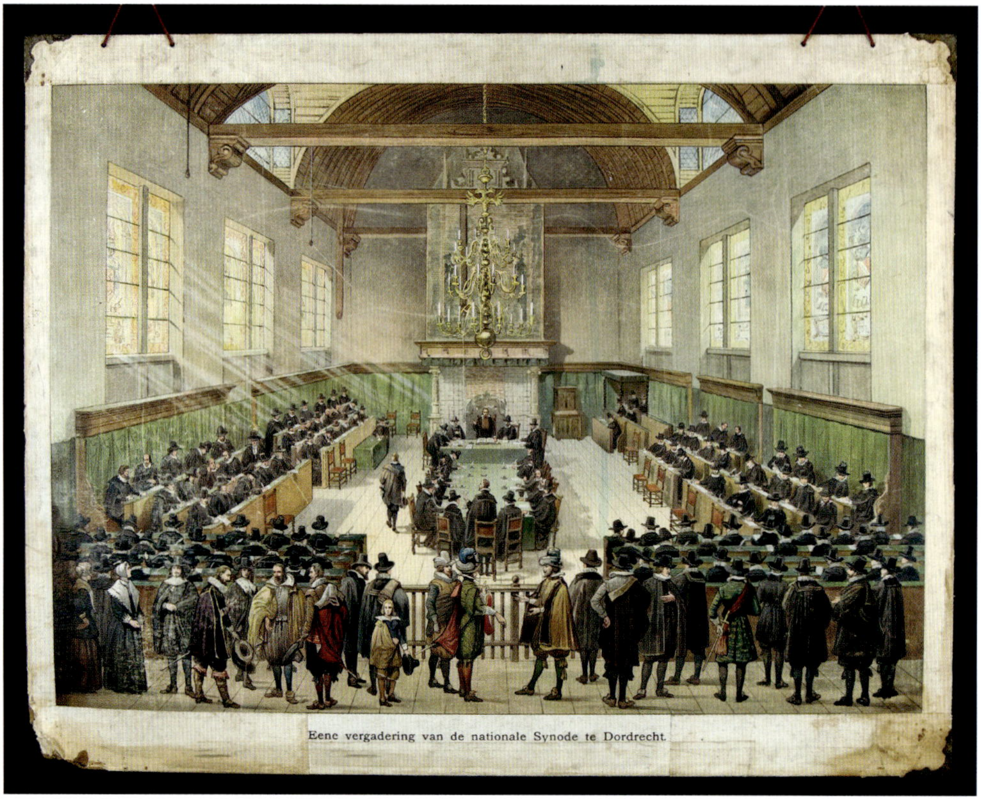

Figure 2 Educational plate of the *The National Synod*, lithograph by G. van Hove. Published by J. B. Wolters, Groningen, 1911.

Due to abrasion and increased transparency of the paint layers, it can be seen that the architectural elements were painted first; not only the large conference room with its wooden floor and ceiling, but also the furniture. In preparation for the application of a green glaze for the upholstery, the furniture was modelled with different shades of grey. In the modelling stage the curtains were given folds, tabletops acquired a shadow and a highlighted edge, and with a darker base colour the whole left side of the conference room was prepared for a more subdued tone of green. Subsequently the still empty tables and desks and the curtains along the walls were covered with a green glaze to imitate green velvet upholstery.

In the next stage all the figures and accessories were applied on top of these paints, and over the green glaze. In accordance with the sunlight, depicted coming through the windows on the left, shadows were added using a second layer of green glaze of a darker shade.

Restoration history

Research into the painting's restoration history revealed archival documents of the last full restoration treatment of 1926 when Hendrik G. Luitwieler, a Rotterdam painter and restorer, removed the varnish and applied a wax-resin lining. Luitwieler records in a letter that in the past the tablecloths and curtains had been overpainted with a 'fresh green' colour.[4] There is a visual record of the overpaint as Luitwieler describes it – in 1911 the painting was reproduced as an educational print (*schoolplaat*) in

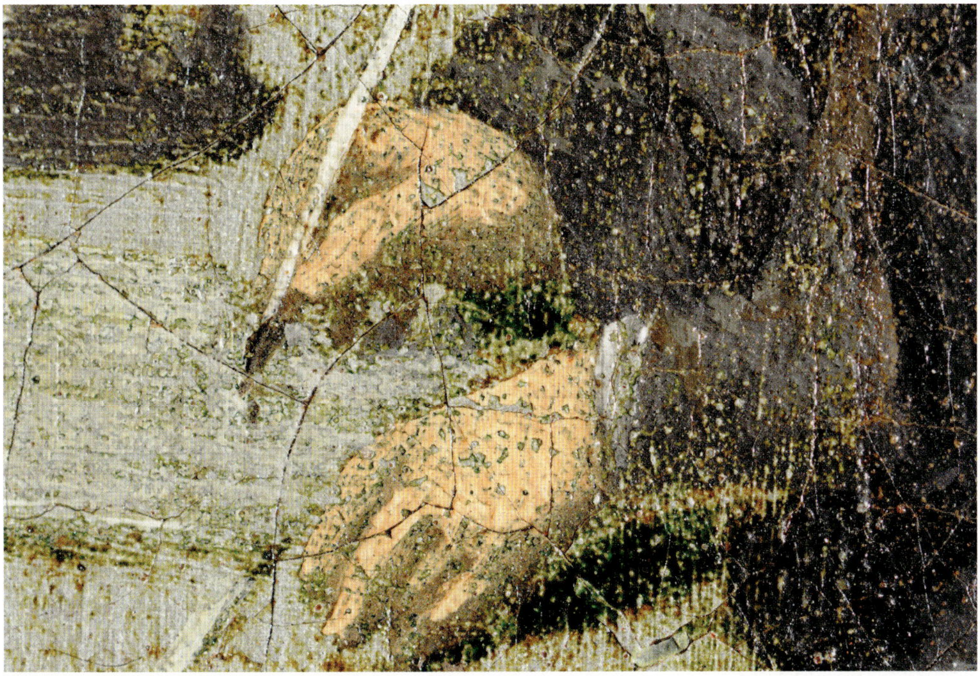

Figure 3 Detail (hands of the top right figure in Figure 5) of the grey underpaint of the tablecloth on the central table with the Arminians. Remnants of (discolored) glaze are visible in the shallows of the brushstrokes and better preserved dark green glaze can be seen as shadow areas under the hands.

a series showing the 24 most important historical events in the Netherlands (see Figure 2). The lithograph translates the painting into a horizontal composition and other artistic liberties have been taken, yet the colour of the upholstery is green throughout; presumably this is how the printmaker would have seen it on the painting.

In 1926 Luitwieler removed this overpaint revealing the grey preparation layers. Apparently the original green glaze had been removed, probably in the 19th century. Small remnants of the original glaze that are still present in the pits and grooves of the paint structure show a brown discoloration (see Figure 3).[5] This colour change must have been the motive for removing the original glaze, which was probably done with solvents since there are no signs of scraping. The green glazes in the shadows are better preserved: in the curtain on the left side of the painting and in the shadows behind figures and objects. In these areas the glazes seem to have retained their original colour, probably thanks to the thickness of the application or because of a different paint composition.

Luitwieler's approach towards the missing green glaze was unassuming and he decided to leave the underlayers visible. As a result the painting had a more subdued palette since the cleaning of 1926, consisting mainly of browns, greys and blacks, with an occasional dash of colour in red or blue, especially in the tapestry over the chimney and some of the elegantly dressed visitors in the forefront (see Figure 4). The archival documents do not provide information on any discussion regarding the approach to treatment between the restorer and the museum director.[6] In his letter from 1926, Luitwieler does, however, show his appreciation for the original, although damaged, paint layers saying that the colour of the underpaint gives the painting more 'style'.

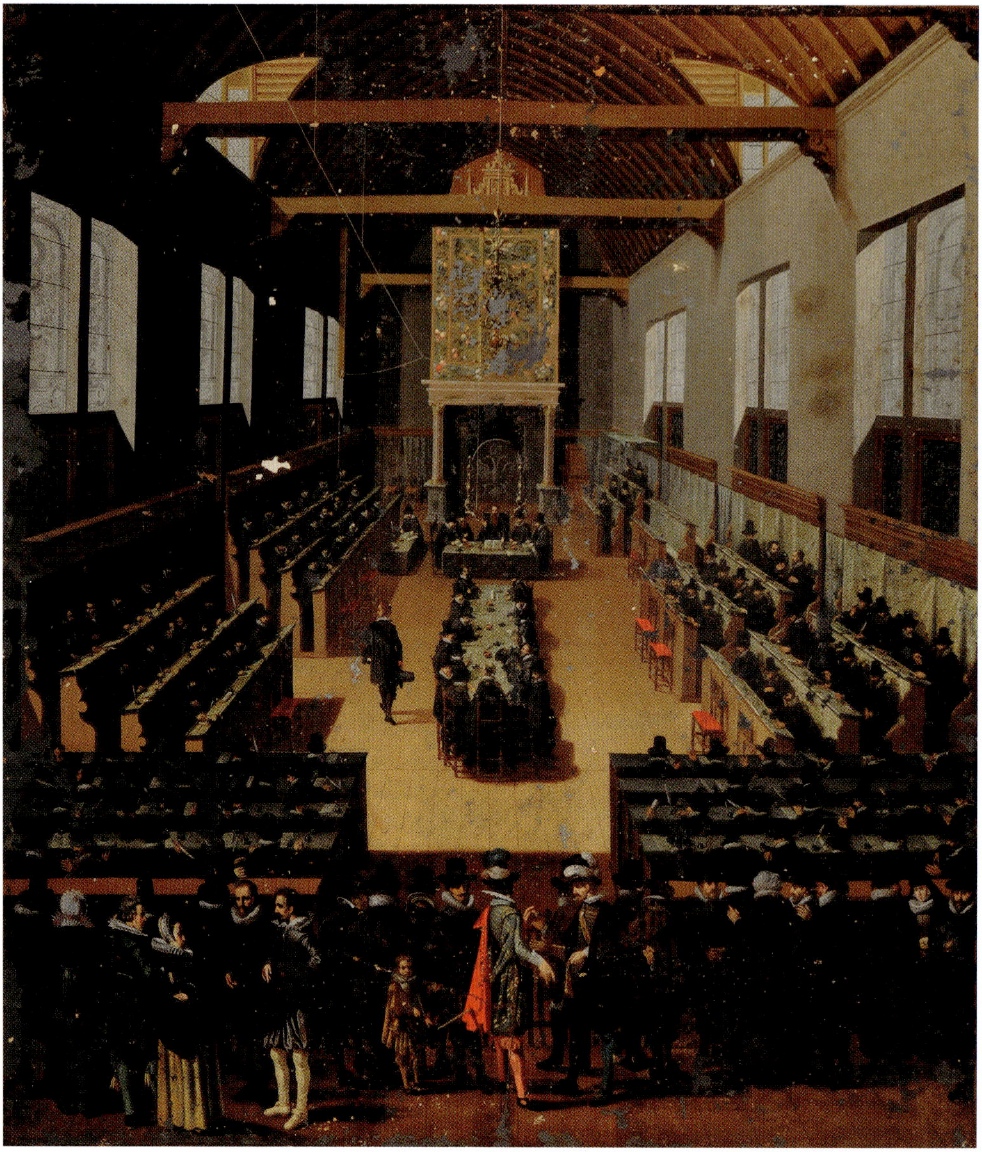

Figure 4 *The National Synod*, during restoration: stripped state.

The present restoration

During the restoration the varnish applied in 1926 was removed together with the retouchings of the same date and older retouchings that were more difficult to dissolve. These older retouchings were placed on top of grey fillings that were much larger than the damages they had to cover. Luitwieler had left these old, hard fills, but we decided to minimise them in order to reveal as much original paint as possible, which meant weeks of careful scraping with up to 200 scalpel blades.

Once the painting had been cleaned the approach to retouching was discussed by a small group consisting of the director of the Dordrecht Museum, the curator, the head of collections and the restorers, and it was decided

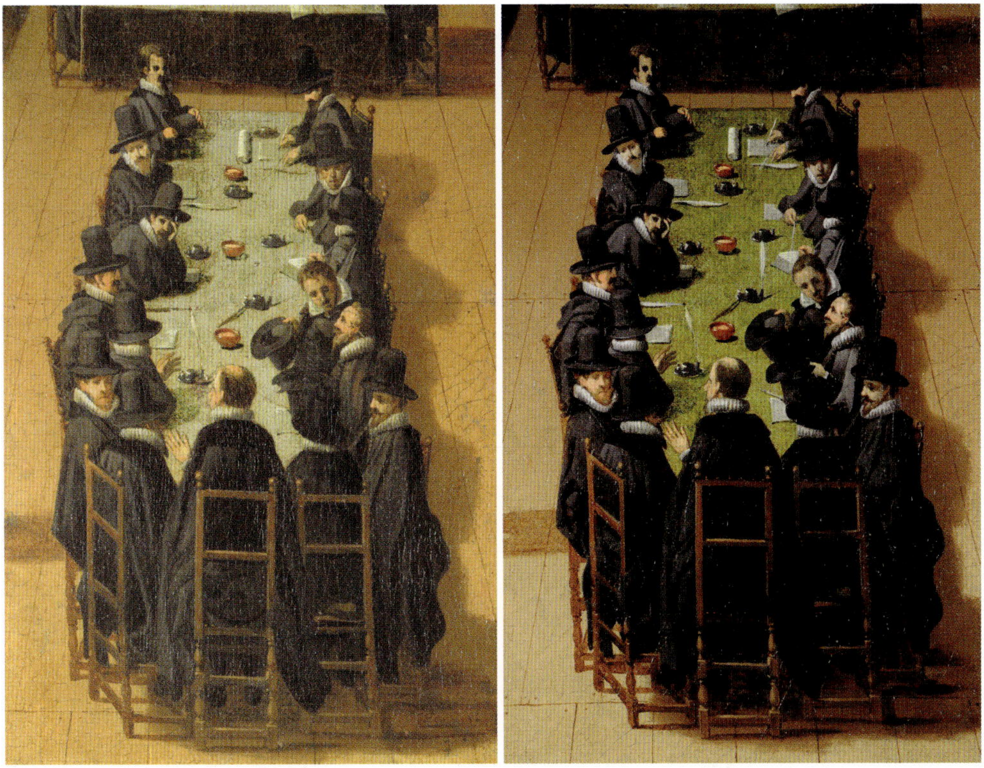

Figure 5 Detail of the table with the Arminians in *The National Synod*, before and after restoration. The details on the table, especially the quills and pieces of paper, stand out more clearly thanks to the reconstruction of the green glaze.

to reconstruct the missing green colour of the textiles. Although overpainting large parts of a painting is normally not done, so as not to obscure original paint layers, we felt that in the case of *The National Synod* it was justified, both artistically and historically.[7]

The choice for a reconstruction entailed overpainting approximately 10% of the surface, therefore the decision was made after careful weighing of the arguments for and against reconstruction. In our opinion the fact that the grey preparation layers are original did not present an argument against reconstruction since they were always meant to be covered by another layer. Moreover, the reconstruction would be applied as a transparent glaze and so the modelling function of the underpaint would be retained. Another argument for the reconstruction was that it would integrate remnants of the old glaze; these islands would otherwise have to be retouched or at least toned down to make them less disturbing. Although a reconstruction would merely be our interpretation of the original appearance, we felt that the decision not to reconstruct could be considered just as subjective. At least the reconstruction would bring the painting closer to its original colour and tonal balance, and it would restore the legibility of the many small details on the tabletops (Figure 5).

Technique of the reconstruction

Prior to the actual reconstruction of the green glaze, two tests were carried out using Gamblin Conservation Colors, mixing Prussian blue and transparent earth yellow. The colour was based on 1.5 cm of original material along the right edge, which still substantially retained its transparent green colour. After discussion with the team the tests were removed. For the actual reconstruction, Mowilith 20 was chosen as the binder, since a Laropal A81 brush varnish was planned, which would have disturbed the Gamblin Conservation Colors. The medium was coloured with Prussian blue, Spritzfarben yellow and a little umber and ochre.

The green glaze was applied over the grey paint layers in a single day. The paint was mixed in a large quantity on a palette and diluted to a glaze-like consistency. The hue of the colour was adjusted to match the original remnants of green present in each section. The glaze had to be painted around all the contours and small details, which required a relatively fine brush, however 'speed' was needed so that the brushstrokes would flow into one another in imitation of an evenly applied glaze. As this kind of overpainting is so alien to our work, it was also important to paint quickly in order to avoid hesitation. Afterwards, the colour was judged to be too strong and bright in some areas. These areas were toned down with a thin glaze consisting of mainly umber. Again speed was needed, so as not to dissolve the first glaze layer.

The result

The visual impact of the green glaze is even greater than we had anticipated: it improves the sense of perspective and light in the painting. Moreover, the details stand out much more clearly and the distribution of bright colours in the painting is more balanced. The green fits the painting so well that unknowing observers, even those familiar with the painting, do not notice the restoration. The choice to reinstate the green colour was the best option for this painting in our opinion. Of course, the bold step of overpainting 10% of the painting's surface could not have been made without the support and enthusiasm of the museum staff.

Notes

1. The restoration of *The National Synod* was made possible through the financial support of Het Hendrik Muller Vaderlandsch Fonds. Johanneke Verhave restored the picture in the Dordrechts Museum conservation studio. The authors are grateful to all contributors to the discussion on the reconstruction of the glaze during the restoration.
2. 'Pouwel Weyts schilder vierentachtich pond(en) over de leverantye van een schilderije vant synodus hange(nde) inde heeren camer opt stadhuys deser stede bij orde en(de) quitan(tie) LXXXIIIJ', Dordrecht Regional Archives, *Thesauriers van de Reparatien*, no. 2995.
3. Material technical research was limited to observations with the naked eye and a stereomicroscope. No cross-sections have been taken, yet the abraded condition of the painting gives a very good insight into the build-up of the paint layers.
4. The letter is kept in the archives of the Vereniging 'Oud-Dordrecht' (The 'Old-Dordrecht' Society), Dordrecht Regional Archives, no. 129-163.
5. Although no analyses were carried out, this seems to point to the discoloration of a green copper glaze. For more information see: van den Berg *et al.* 2000; van Eikema Hommes 2004.
6. See note 4 above. In the letter Luitwieler invites the director to his studio to come and see the painting during the cleaning. No record of such a visit has been found.
7. For another reconstruction of a green glaze see Cove 2010: 81–2.

References

Cove, S. (2010) 'Retouching with PVA resin medium', in R. Ellison, P. Smithen and R. Turnbull (eds), *Mixing and Matching: Approaches to Retouching Paintings*. London: Archetype Publications, 74–86.

Marijnissen, P. *et al.* (eds) (1992) *De Zichtbaere Werelt; schilderkunst uit de Gouden Eeuw in Hollands oudste stad*, exh. cat. Zwolle: Waanders.

van den Berg, K.J., van Eikema Hommes, M., Groen, K. *et al.* (2000) 'On copper green glazes in paintings', in J. Goupy and J-P. Mohen (eds), *Art et chimie, la couleur: Actes du Congrè.* Paris: CNRS Editions, 18–21.

van Eikema Hommes, M. (2004) 'Verdrigris glazes in historical oil paintings: recipes and techniques', in *Changing Pictures: Discoloration in 15th–17th-Century Oil Paintings.* London: Archetype Publications, 51–89.

Author's addresses

- Johanneke Verhave, Gordelweg 191a, 3038GH Rotterdam, The Netherlands (johanneke@restauratieatelierrotterdam.nl)
- Lidwien Speleers, Dordrechts Museum, Dordrecht, The Netherlands (lbj.speleers@dordrecht.nl)

Vivid colours in the Golden Room: an educated but tentative reconstruction of an ensemble by Pellegrini in the Mauritshuis

Carol Pottasch, Sabrina Meloni, Ivo Hoekstra, Margriet van Eikema Hommes and Ralph Haswell

Introduction

This paper presents a study of the original colours of the paintings in the Golden Room of the Mauritshuis. The ensemble, painted by the Italian artist Giovanni Antonio Pellegrini in 1718, is one of the few early 18th-century ensembles that has survived *in situ* in the Netherlands. While the Mauritshuis was closed for renovation, the opportunity was taken to examine and treat these paintings. This study is based on research into the painting technique and the analyses performed on cross-sections taken from this group of paintings. The composition of the pigments was analysed with scanning electron microscopy-energy dispersive X-ray (SEM-EDX) analysis in collaboration with the Shell Technology Centre Amsterdam. Unfortunately, the fading and degradation of some of the original colours in the paintings has given them a completely different appearance, and as a result the concept of unity between the paintings and the architecture has lost much of its original presentation. In order to gain a better understanding of the original appearance of the scheme, an educated but tentative Photoshop reconstruction was made. Based on paint sample analyses of the different elements in the ensemble, the reconstruction gives an impression of the joyous effect of the original light and bright colouring (Figure 1).

Pellegrini

Giovanni Antonio Pellegrini (1675–1741) was a Venetian painter who travelled and worked extensively throughout Europe. He stood at the forefront of a new, rapidly developing and highly popular Venetian aesthetic and style, pioneered by Sebastiano Ricci (1659–1734) and his works show the influence of Rubens (1577–1640), Veronese (1528–1588) and Luca Giordano (1634–1705). In turn, his paintings strongly influenced Tiepolo, and in the Netherlands, Jacob de Witt. Pellegrini painted many decorative cycles with allegorical subjects and themes from classical mythology. He was highly regarded and received many commissions in England and Germany.[1]

The Golden Room

After being damaged by a fire in 1704, the interior of the Mauritshuis was destroyed.[2]

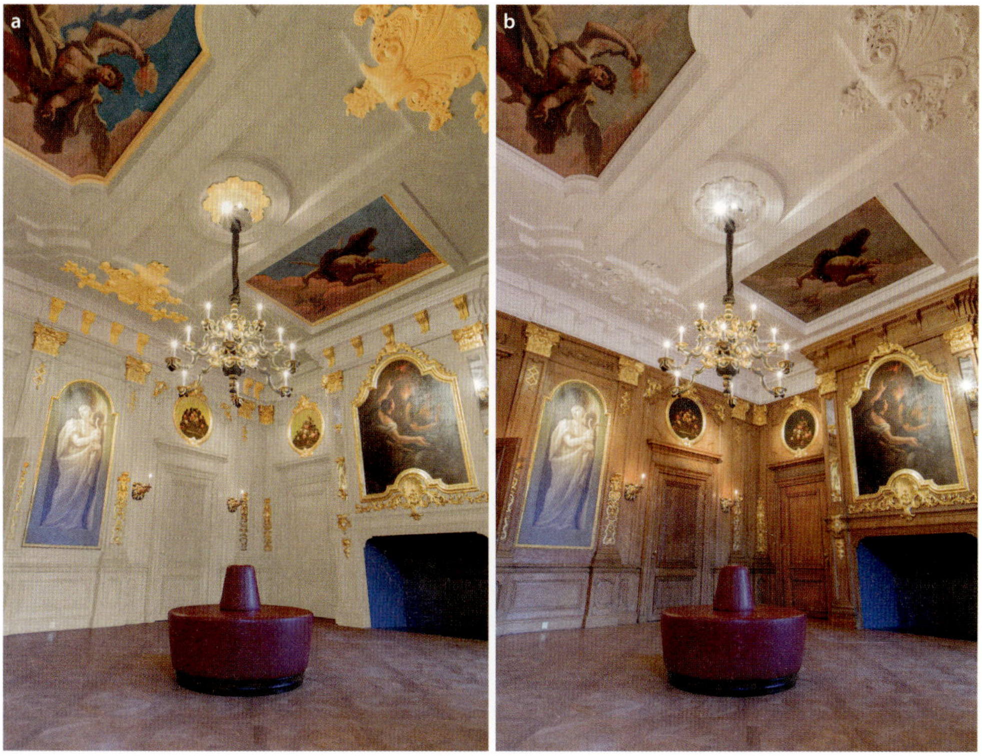

Figure 1 (a) A reconstruction of one corner of the Golden Room giving an impression of *The Dwindling Night* with a vivid blue sky and warm pink clouds, flower still lifes with a gold-coloured background and gold-coloured decorations. Only the ceiling should be in the same dark colour. (b) The same corner of the Golden Room at present showing the bare, paint-stripped panelling, the flower *tondos* without the gold-coloured background and the discoloured ceiling painting, *The Dwindling Night*.

In 1718, Pellegrini was asked to redecorate the Golden Room in the Mauritshuis with 15 paintings. Today 12 of these painting are set into plain oak wood panelling which covers the walls from floor to ceiling: four (slightly darkened) *grisailles* representing the four elements adorn the oak panelled walls, six flower *tondi* (round paintings which nowadays appear quite dark) can be seen above doors, and two chimney pieces with allegorical scenes hang above the fireplaces. The now white ceiling is punctuated with three ceiling paintings representing the gods *Aurora*, *Apollo* and Nox in *The Dwindling Night*. Each of these paintings, in which a classical god is depicted floating in the sky, is surrounded by a white frame. The magnificent *riche* room is further adorned with white acanthus leaves on the ceiling, gilded frames around the paintings on the walls, as well as small gilded decorative elements.

Architectural paint research has indicated that the once light, elegant and decorative scheme of the early 18th century has lost much of its splendour due to the removal of all the architectural paint from the oak wall panelling and the overpainting of the wooden ceiling (van Eikema Hommes *et al.* 2014). It could be established that, in the original scheme, both the wall panelling and ceiling were first painted in a light grey colour (possibly in imitation of grey stone). Traces of this paint were discovered at wall level in 'hidden' areas that escaped paint removal in the 19th century and on the ceiling under thick layers of overpaint. On top of this grey layer, the mouldings and decorative

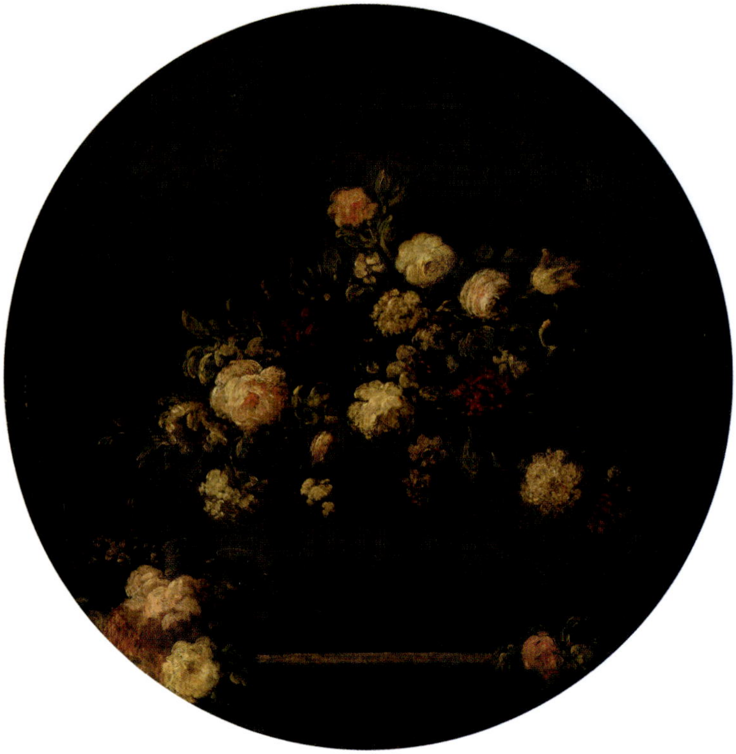

Figure 2 *Flower Still Life*, 1718, oil on canvas, 96 cm Ø, inv. no. 1145, The Hague, Mauritshuis (photo: Margaretha Svenson).

wood carving on the walls were gilded. The moulding and carvings of the ceiling, the capitals and the consoles also had a gold-coloured finish, but one with brass leaf.[3]

Flower *tondi*

During the thorough examination prior to the conservation treatment of the six flower *tondi* (Figure 2), which are located above the doors in the room, it became clear that the present appearance does not reflect how they looked in the 18th century (Meloni *et al.* 2014). While most of the paintings of the Golden Room underwent full conservation treatment, the round flower paintings only received minor aesthetic treatment: surface dirt was removed and a thin layer of varnish was applied, mainly because all the backgrounds had been heavily overpainted in a brown-green tone and very little original paint remained beneath this layer. Traces of the original background could only be seen along the perimeter of the painting and in some losses. Under the stereomicroscope a glittering gold-coloured material was observed in all six *tondi*. In some cases a translucent green layer was seen above and below the golden material. In cross-sections, a metal leaf layer was visible beneath the paint layers. When this was analysed with SEM-EDX, it became clear that the gold-coloured metal leaf was not real gold: the elements copper (Cu) and zinc (Zn) were detected in both the metal leaf and the translucent green areas around the metal leaf. It was determined that instead of gold leaf, much cheaper brass leaf had been used. Over time, the brass reacted with the oil from the paint layers above and below it, forming a

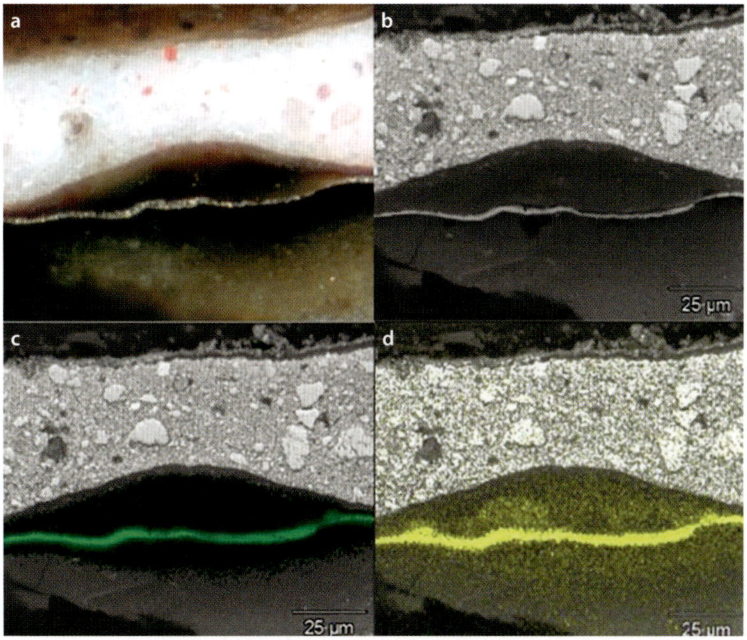

Figure 3 (a) Cross-section MH1145x2b taken from the pink flower in the middle: the brass layer and the green degradation of the brass can be detected; (b) backscatter image; (c) elemental mapping of the copper (Cu) present; (d) elemental mapping of the zinc (Zn) present.

translucent green layer. In some cross-sections the metal leaf can still be detected, but in others only the green degradation product is visible and none of the metal leaf remains (Figure 3).

Using infrared imaging, it was possible to detect roughly square shapes from the individual sheets of brass leaf throughout the composition, suggesting that the whole canvas was covered in a layer of brass leaf prior to painting. A layer of yellow mordant was applied on top of the ground layer to adhere the brass leaf. The flowers and the stone slabs were subsequently painted over the gold-coloured background. Gold-coloured metal leaf backgrounds were not unusual for flower overdoors in the late 17th and early 18th century. In Huys Schuylenburg (1715) in the Hague, which is located closed to the Mauritshuis, several flower overdoors with gold-coloured metal backgrounds are still *in situ* in its interior. As the backgrounds are well preserved it seems likely that real gold was used.[4] These examples give an idea as to how the six flower *tondi* in the Golden Room may have first appeared. It was suggested that in the stone architecture, the mouldings and decorative carving were made of gold, while the flower *tondi* created the illusion that above each door a pot with flowers was placed before a golden background. The use of the cheaper and less stable brass leaf was probably for economic rather than aesthetic reasons as it was about the tenth of the price of gold leaf.

Bright blue and pink

The Dwindling Night provides an example of how the appearance of the ceiling paintings has changed over time (Figure 4). Along the edges of the sky that were covered by the frame, areas of vivid blue, pink and bright orange paint were discovered. Cross-sections were taken from these areas to analyse the build-up of the paint layers and use of pigments. The sky was painted with combinations of smalt and red

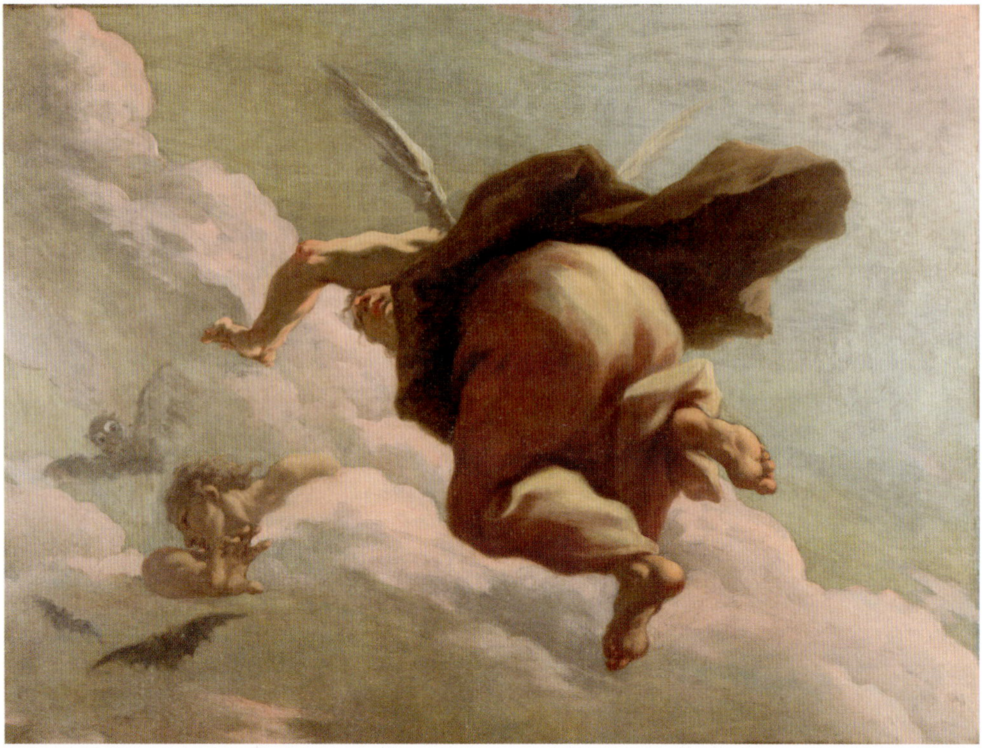

Figure 4 *The Dwindling Night*, 1718, oil on canvas, 192 × 262 cm, inv. no. 1137, The Hague, Mauritshuis (photo: Margaretha Svenson).

lake, which have faded to duller tones of blue and pink. Cross-sections taken from the vivid blue area, preserved at the covered edge, show blue smalt particles in bright field (Figure 5a). EDX analysis indicated that the smalt particles contain a high content of potassium and cobalt, and the surrounding paint layer contains lead white with a small amount of earth pigments (containing aluminium and iron) (Figure 5b).

In a cross-section taken from the discoloured area in the central field of the painting, only greyish-blue degraded smalt particles were detected in visible light (Figure 5c). EDX analyses confirmed that these particles have a low potassium content. The mapping of an area in this cross-section showed the degradation of the smalt: all the potassium has leached out of the glass particles into the binding medium (Spring *et al.* 2005) (Figure 5d). However there appears to be no difference in the cobalt content of the smalt particles in samples from the faded and unfaded areas. As with the samples from the unfaded edges, only lead white and a trace of earth pigments (EDX: alum and iron) were present in addition to the smalt, therefore it seems safe to assume that the faded paint would originally have been more or less the same colour as the unfaded edges.

In the top right corner of the painting, a bright orange-pink was found to be well preserved under the frame. This provided an insight into how large areas of the pink-orange clouds may have once appeared. EDX analyses of a cross-section taken from a pinkish area showed that the layer contained a mixture of lead white and vermilion. Furthermore, analyses of this cross-section indicated the presence of aluminium. In EDX mapping of an area of this cross-section, aluminium was detected

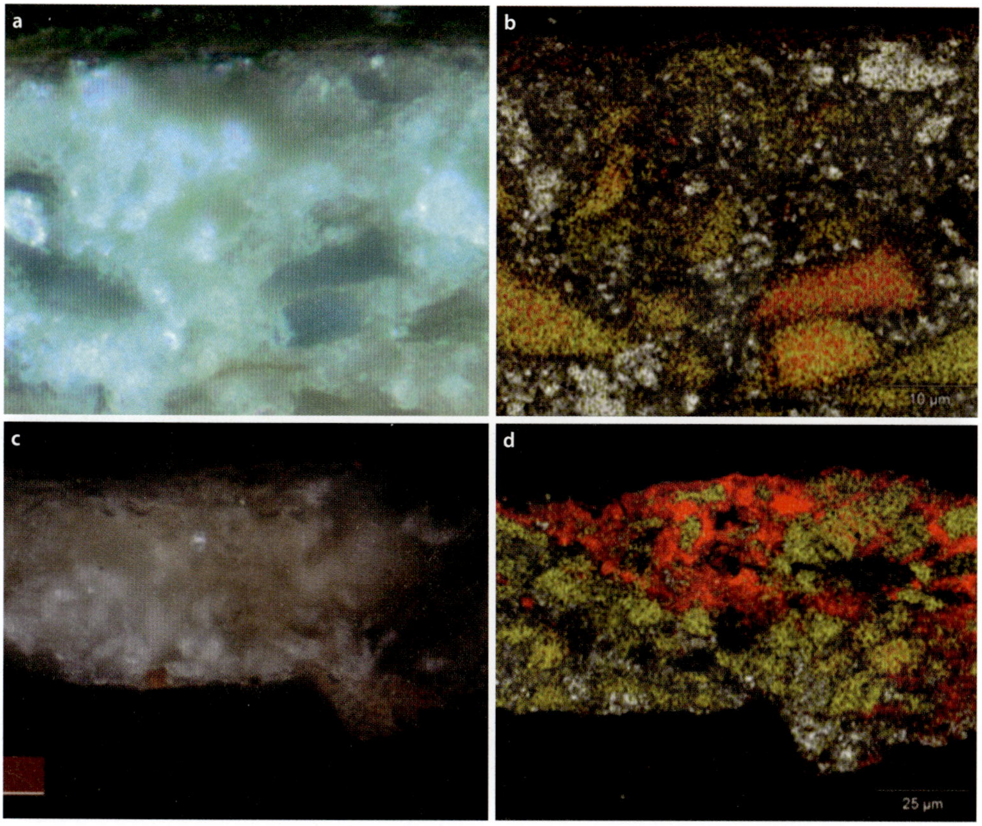

Figure 5 (a) Cross-section taken in the not/less degraded blue area in the sky (bright field, ×40): blue particles relating to smalt can be identified; (b) in the backscatter image (×1000) potassium and cobalt (red) can be identified in the particles; (c) cross-section taken from a severely degraded bluish area in the sky (dark field, ×50): blue particles relating to smalt can only be identified in the bottom of the layer; (d) in the backscatter image (×500) cobalt (red) can be identified in the particles; the potassium however has leached into the oil medium.

throughout. It seems very likely that the aluminium relates to an aluminium-containing substrate for a lake pigment. As the unfaded area of the painting appears pink-orange in colour, and the paint contains vermilion, it is probable that this was a yellow lake, which has since faded.

Sculptures in paint

Over the years the *grisailles* representing the four elements have changed more in colour and tone than is obvious at first glance. Although it would seem logical that all of the canvas for Pellegrini's ensemble would be ordered at the same time, and that the composition of the ground of the *grisailles* and the other paintings would be identical, this is not the case: while the *grisailles* have an almost pure white ground, the other paintings have a grey-brown ground that contains lead white, earth colours, quartz, bone black and chalk (Pottasch *et al.* 2015).

The difference in the ground layers may relate to the fact that the artist sought to achieve the cool tone of light marble statues from the first paint layer. It seems that Pellegrini used a very economical way of painting for the *grisailles*. Firstly, the layers are all very thin and limited in number and secondly, it is noteworthy that the order of the ground layers is unusual: the first layer consists almost exclusively of lead white, while the second contains a mixture of lead white and chalk. A trace amount of bone black was added to give these layers a cooler appearance. The final paint layer also consists mainly of lead white with small amounts of bone black and yellow earth.

Over time, the extremely thin paint layers have become more transparent so that the underlying canvas support is now much more visible. The darkening of the oil binding medium and the fact that the paintings have been lined with wax-resin adhesive must also have contributed to the warmer and darker appearance of the pictures. Examination of cross-sections indicated that at first the tone of the paintings would have been much cooler and closer to the original colour of the walls, creating the illusion of a stone statue in a niche.

While the *grisailles* were once in unity with the grey walls, the ceiling paintings would have given the impression of skylights. With the whole scheme together, Pellegrini must have aimed to create the illusion of antique architecture of a grey stone pavilion, with cool marble statues set into stone niches and openings in the ceiling giving a view to a colourful sky. In the Photoshop impression, these aspects and the rich effect of the flower *tondi* can be seen (see Figure 1a).

Acknowledgements

The authors would like to thank Susan Smelt, Julia van den Burg, Laura Hartman, Oleg Karuvits, Claire Toussat, Petria Noble, Annelies van Loon, Bob van Wingerden, Fred Meijer, Margareta Svensson, Ige Verslype, Arie Pappot, Roos Keppler, Eva Berger, Abbie Vandivere and the American Friends of the Mauritshuis. The research was carried out within the framework of the Partners in Science collaboration between the Mauritshuis Conservation Department and Shell Technology Centre Amsterdam.

Notes

1. For information on Pellegrini see: Aikema and Mijnlieff 1993; Knox 1995.
2. The house was originally built between 1630 and 1640 by Johann Maurits von Nassau Siegen and was designed by Jacob van Campen.
3. However, the metal leaf appeared to be completely degraded: in cross-sections only a thick, dark medium-rich layer was found. SEM-EDX indentified only copper and a trace amount of zinc. At first this suggested that copper leaf had been used and that the zinc possibly originated from the later zinc white layers applied on top (van Eikema Hommes *et al.* 2014). However, further investigation identified a fragment of preserved metal leaf that contained a low amount of zinc indicating that brass leaf had been used for these elements. Our initial misunderstanding arose because of the fact that when the brass corroded, the copper remained in place in the paint system while the more reactive zinc had almost completely disintegrated.
4. At Huis Schuylenburg in The Hague, there is a *sopraporte* of flowers, oil on canvas, early 18th century, attributed to Caspar Pieter Verbrugghen II (1664–1730) with a golden background: see van Zoest and van Eck 1988: 79, ill. 89.

References

Aikema, B. and Mijnlieff, E. (1993) 'Giovanni Antonio Pellegrini, 1716–1718: a Venetian painter in the Low Countries', in J. de Jong, D. Meijers, H. Pennock and V. Schmidt (eds), *Nederland-Italie: Artistic Relations between the Low Countries and Italy 1400–1750, Netherlands Yearbook for the History of Art*, vol. 44. Zwolle: Waanders Uitgevers, 215–42.

Knox, G. (1995) *Antonio Pellegrini 1675–1741.* Oxford: Clarendon Press.

Meloni, S., Salazar-Walsh, M., Haswell, R. and Toussat, C. (2014) 'Golden paintings for the Golden Room of the Mauritshuis? The search for the original appearance of six flower tondos', poster presented at *Technology and Practice: Studying 18th-Century Paintings and Art on Paper, Copenhagen, 2–3 June 2014*. Available at: www.cats-cons.dk/wp-content/uploads/2014/07/Meloni_Cats_Poster.pdf (accessed 22 March 2016).

Pottasch, C., Smelt, S. and Haswell, R. (2015) 'Breaking new ground: investigating Pellegrini's use of ground in the Golden Room of the Mauritshuis', in H. Evans and K. Muir (eds), *Studying 18th-Century Paintings and Works of Art on Paper*, CATS Proceedings II. London: Archetype Publications, 20–23.

Spring, M., Higgitt, C. and Saunders, D. (2005) 'Investigation of pigment-medium interaction processes in oil paint containing degraded smalt', *National Gallery Technical Bulletin* 26: 56–70.

van Eikema Hommes, M., Keune, K., Jongsma, R., Polman, M. and Pottasch, C. (2014) 'The original colour finish of the Golden Room', in Q. Buvelot (ed.), *Mauritshuis: The Building*. Zwolle: Waanders Uitgevers, 152–7.

van Zoest, R. and van Eck, X. (1988) *Huis Schuylenburch*. Den Haag: SDU Uitgevers.

Authors' addresses

- *Author for correspondence*: Carol Pottasch, Senior Paintings Conservator, Royal Cabinet of Painting The Mauritshuis, The Hague, The Netherlands (c.pottasch@mauritshuis.nl)
- Sabrina Meloni, Paintings Conservator, Royal Cabinet of Painting The Mauritshuis, The Hague, The Netherlands
- Ivo Hoekstra, Multimedia Producer, Royal Cabinet of Painting The Mauritshuis, The Hague, The Netherlands
- Margriet van Eikema Hommes, Researcher Cultural Heritage Agency and Associate Professor, Delft University of Technology, Delft, The Netherlands
- Ralph Haswell, Principal Scientist, Shell Technology Centre Amsterdam, The Netherlands

Investigating surface phenomena in paints and paintings by Edvard Munch (1863–1944) in the context of the artist's painting materials

Jin Strand Ferrer, Alyssa Hull and Terje Syversen

Introduction

When Edvard Munch's will was read after the artist's death in 1944, the city of Oslo in Norway was the main beneficiary: Munch had donated a major part of his own extensive art collection and inventory of his estate (Figure 1). This generous gift left the city of Oslo with a collection which today consists of approximately 1,150 paintings, 25,500 artworks on paper and a large collection of art materials and supplies: 897 paint tubes, palettes, paintbrushes, bottles of binding medium, chalk, pastel, crayons and pencils. The major task of registering and cataloguing this inventory was initiated in 1945 together with the first conservation campaign (Langaard 1951: 75). This work lasted until the opening of the Munch Museum in 1963. Many of the artworks bequeathed were already in poor condition, partly because of the experimental techniques and materials used, but also due to lack of suitable storage facilities and the artist's own negligence, which was much commented on by his friends while the collection was in Munch's care (Dørje-Haug 1949).

Research in the Munch Museum's Conservation Department has often focused on exploring the variety of Munch's painting techniques (Dørje-Haug 1949; Thurmann-Moe 1995; Topalova-Casadiego 2009; Singer *et al.* 2010; Plahter and Topalova-Casadiego 2011; Kutzke and Topalova-Casadiego 2012), and more recently also on issues related to the degradation of materials (Sandbakken and Tveit 2012; Nasa *et al.* 2015). In 2014, two research projects were initiated to assess the materials and their deterioration with the aim of developing individual approaches to conservation treatments:

- The main purpose of the 'Munch's materials and paint tubes' research project is to register and catalogue the collection of artists' materials and to analyse the contents of the paint. It will also investigate possible decay and/or changes to binder(s), pigments and fillers.
- The project 'Investigating colour changes in Munch's paintings' will assess colour changes in paintings to inform exhibition policies and storage solutions.

Munch's materials and paint tubes

During the period 2009–2011, the museum collaborated with the Cultural History Museum in Oslo to analyse Munch's paint. This work first published results from a small selection of samples (Kutzke and Topalova-Casadiego

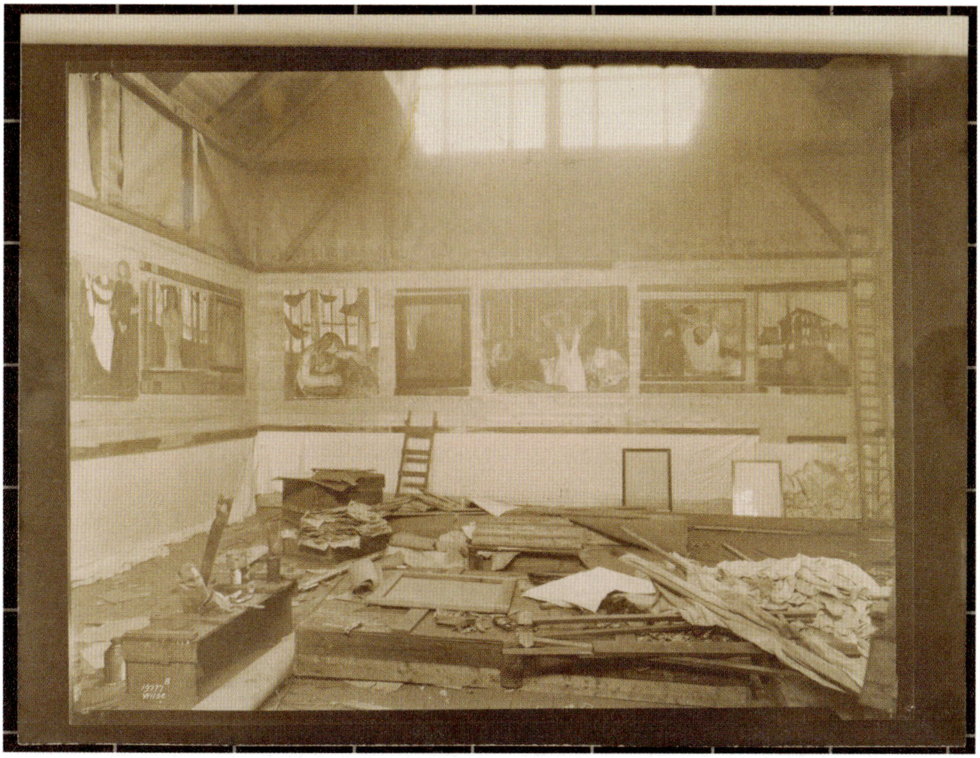

Figure 1 Munch's studio at Ekely in Oslo, Norway. (Photo: Anders Beer Wilse, 1925)

2012) and is now being developed further following the museum's acquisition of analytical equipment.[1]

The first part of this project involved cataloguing all existing photographic and written documentation. Sampling and documentation of each item has also been carried out. In instances where there are several items of the same brand and pigment, only one sample has been taken. The museum's collection consists of various brands, several of which are no longer in production.[2] Analysis has been performed on binding media, pigments and additives from several paint tubes and the results of this will be published in a future article.[3] The obvious variations in binding media have a particular effect on the paints' physical properties, and this is evident while holding and gently squeezing the different tubes: many have dried out and become hard, some even brittle, while others are still completely liquid. Most, however, are somewhere between these two extremes (Figure 2).

Our research has shown that Munch's paint tubes are likely to have been produced over a long time span, between 1864 until after 1930 – a period when an extensive new range of pigments and materials were introduced to the industry (Carlyle 1999; Clarke 2009; Townsend *et al.* 1995). The results from the aforementioned analysis could yield important information regarding questions about authenticity and dating.

Investigating colour changes in Munch's paintings

Irregular colour variations have been observed in monochrome brushstrokes in some of Munch's paintings. The phenomena observed,

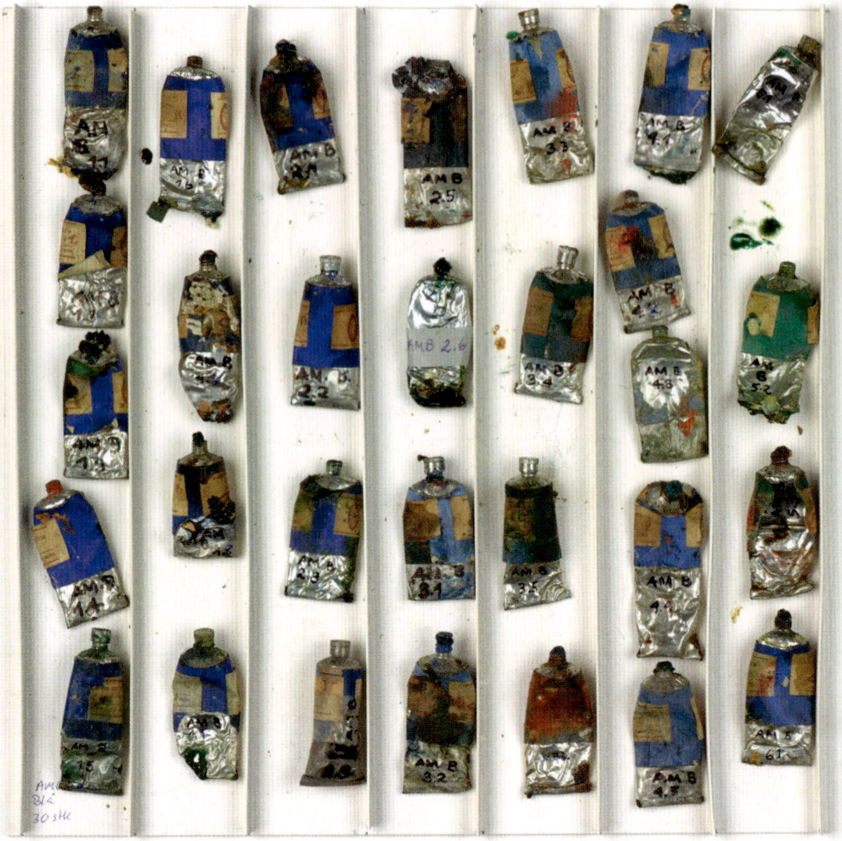

Figure 2 A selection of paint tubes from the Munch Museum's collection of the artist's painting materials (© The Munch Museum).

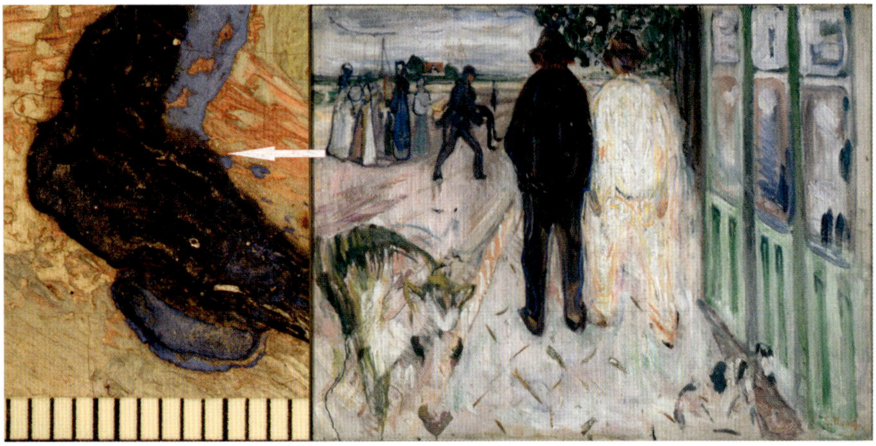

Figure 3 Detail from *The Drowned Boy*, from the painted outline of the female figure's dress in the background, far left. An example of the phenomenon observed in connection with cobalt blue in Munch's paintings (© The Munch Museum).

which include darkening and bleaching of blue and red painted areas, could be mistaken for a wet-in-wet paint application, but visual investigations have shown alterations on the surfaces (Figure 3). Preliminary analysis with portable X-ray fluorescence spectroscopy (pXRF) indicated that the pigments associated with these are cobalt blue and synthetic ultramarine as well as red lakes. The focus of this study is on paintings from 1907 to 1908, a period when Edvard Munch was living in Warnemünde, a seaside village in Germany, as paintings from this period seem to have more frequent occurrences of these phenomena. In the museum's collection there are 49 paintings from the Warnemünde period, among them *The Death of Marat*, from the *Green Room* series, *Bathing Men*, portraits and village scenes from Warnemünde (Bardon *et al.* 1999).

The first part of this investigation aims to identify and characterise the pigments and paint materials where alterations are observed. A comparative study with existing research into Munch's contemporaries will also be carried out. Munch was an innovative and productive artist, known for experimenting with different painting materials and binding media. He used a variety of painting techniques, often within the same painting. Analysis with gas chromatography-mass spectrometry (GC-MS) has detected the presence of natural wax and plant resins in addition to drying oils (Nasa *et al.* 2015). Animal glue, casein and egg have also been identified in samples collected from some of Munch's paintings (Singer *et al.* 2010). Fourier transform infrared spectroscopy-attenuated total reflectance (FTIR-ATR) analysis carried out on some of Munch's paint tubes indicates the presence of wax and/or resins in addition to drying oil. This correlates with technical literature and investigations into other artists' paint and materials from the same period (van de Laar and Burnstock 1997; Carlyle 2001). These materials have also been also indicated by FTIR-ATR analysis of one of the samples from Munch's painting *The Drowned Boy* (Figure 4). Artificial ageing will also be carried out to study degradation mechanisms in samples from Munch's paint tubes by means of spectroscopic techniques.

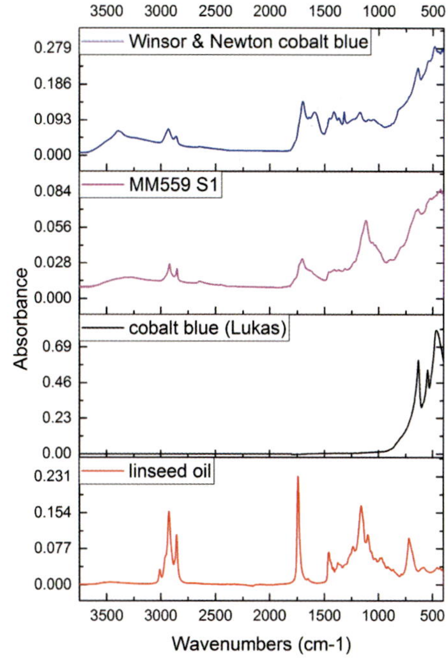

Figure 4 IR spectra collected in ATR mode suggests that the binder in a sample taken from the painting *The Drowned Boy* and from commercial paint formulations may not consist of drying oils alone (© The Munch Museum).

Conclusions

These existing areas of research will contribute to our knowledge of the materials used by Munch and the degradation mechanisms affecting the collection. The main objective is to develop a better conservation strategy and implement sustainable exhibition policies to safeguard Munch's unique legacy.

Acknowledgements

We would like to thank colleagues at the Munch Museum for support and advice.

Notes

1. Thermo Fisher Nicolet bench top spectrometer with an integrated FTIR-ATR, Raman-IR and GC-IR.
2. Alf Bjercke, Arnbak, Sadolin, Blockx, C. Kreul, Devoe & Reynoldsm, Gee Wee, Zet Farbe, Pelikan, Beckers, LeFranc, Lusk & Holtz, Mussini ölfarbe, Rembrandt, Richard Wurm, Reeves-Students, Schoenfeldt, Vilhelm Pacht, Winsor & Newton, Landseer and Newmans.
3. Preliminary results will be published in an article in preparation by Alyssa Hull and Terje Syversen.

References

Bardon, A., Eggum, A., Huusko, T. and Woll, G. (1999) *Munch og Warnemünde 1907–1908*. Drammen: Labyrinth Press.

Carlyle, L. (1999) 'Paint driers discussed in 19th-century British oil painting manuals', *Journal of the American Institute for Conservation* 38(1):69–82.

Carlyle, L. (2001) *The Artist's Assistant: Oil Painting Instruction Manuals and Handbooks in Britain, 1800–1900, with Reference to Selected 18th-century Sources*. London: Archetype Publications.

Clarke, M. (2009) 'A nineteenth-century colourman's terminology', *Studies in Conservation* 54(3): 160–69.

Dørje-Haug, O. (1949) Letter addressed to Johan Langaard, the Director of the National Gallery in Oslo.

Kutzke, H. and Topalova-Casadiego, B. (2012) 'Exploring an artist's practice: Edvard Munch's paint tubes', in S. Eyb-Green, J.H. Townsend, M. Clarke, J. Nadolny and S. Kroustallis (eds), *The Artist's Process: Technology and Interpretation*. London: Archetype Publications, 172–5.

Langaard, J.H. (1951) 'Beretning om virksomheten i femårs-perioden 1946–1951', in Årbok 1946–1951. Oslo: Det Mallingske Boktrykkeri, 60–100.

Nasa, J. La, Zanaboni, M., Uldanck, D., Degano, I., Modugno, F., Kutzke, H., Tveit, E.S., Topalova-Casadiego, B. and Colombini, M.P. (2015) 'Novel application of liquid chromatography/mass spectrometry for the characterization of drying oils in art: elucidation on the composition of original paint materials used by Edvard Munch (1863–1944)', *Chimica Acta* [in press].

Plahter, U. and Topalova-Casadiego, B. (2011) '*The Scream* by Edvard Munch: painting techniques and colouring materials', in M. Spring (ed.), *Studying Old Master Paintings: Technology and Practice*. London: Archetype Publications, 244–52.

Sandbakken, E.G. and Tveit, E.S. (2012) 'Edvard Munch's monumental sketches (1909–1916) for the Aula of Oslo University, Norway: conservation issues and treatments', *Studies in Conservation* 57(1): 258–67.

Singer, B., Aslaksby, T., Topalova-Casadiego, B. and Tveit, E.S. (2010) 'Investigation of materials used by Edvard Munch', *Studies in Conservation* 55: 1–19.

Thurmann-Moe, J. (1995) *Edvard Munchs 'hestekur': eksperimenter med teknikk og materialer*. Oslo: Munch-museet.

Topalova-Casadiego, B. (2009) 'Technical aspects of Edvard Munch's paintings', in G. Woll (ed.), *Edvard Munch Compelete Paintings*, catalogue raisonné (4 vols), vol. 2. London: Thames & Hudson, 425–57.

Townsend, J.H., Carlyle, L., Khandekar, N. and Woodcock, S. (1995) 'Later nineteenth century pigments: evidence for additions and substitutions', *The Conservator* 19: 65–78.

van de Laar, M. and Burnstock, A. (1997) '"With paint from Claus & Fritz": a study of an Amsterdam painting materials firm (1841–1931)', *Journal of the American Institute for Conservation* 36(1):1–16.

Authors' addresses

- Jin Strand Ferrer, Conservation Department, Munch Museum, Oslo, Norway (jin.ferrer@munchmuseet.no)
- Alyssa Hull, Duke University, Raleigh-Durham, North Carolina, USA (Alyssa.hull@fulbrightmail.org)
- Terje Syversen, Conservation Department, Munch Museum, Oslo, Norway (terje.syversen@munchmuseet.no)